Brittle Stars, Sea Urchins and Feather Stars
of British Columbia, Southeast Alaska and Puget Sound

BRITTLE STARS

SEA URCHINS AND FEATHER STARS
OF BRITISH COLUMBIA, SOUTHEAST ALASKA
AND PUGET SOUND

PHILIP LAMBERT AND WILLIAM C. AUSTIN

ROYAL **BC** MUSEUM
Victoria, Canada

Published by the Royal British Columbia Museum, 675 Belleville Street, Victoria, British Columbia, V8W 9W2, Canada.

Printed in Canada.

The front cover features a Basket Star (*Gorgonocephalus eucnemis*) and the back cover shows a cluster of Purple Sea Urchins (*Strongylocentrotus purpuratus*).

Page 144 contains credits and copyright information for all photographs and illustrations shown in this book.

Library and Archives Canada Cataloguing in Publication Data
Lambert, Philip, 1945-
    Brittle stars, sea urchins and feather stars of British Columbia, Southeast Alaska and Puget Sound.

    (Royal BC Museum handbook)

    Includes bibliographical references and index.
    ISBN 978-07726-5618-6

    1. Ophiuroidea – British Columbia. 2. Ophiuroidea – Washington (State) – Puget Sound. 3. Ophiuroidea – Alaska. 4. Sea urchins – British Columbia. 5. Sea urchins – Washington (State) – Puget Sound. 6. Sea urchins – Alaska. 7. Feather stars – British Columbia. 8. Feather stars – Washington (State) – Puget Sound. 9. Feather stars – Alaska. I. Austin, W. C. (William Carey), 1936– II. Royal BC Museum.

QL384.O6L35 2006   593.9'409711   C2006-960174-7

CONTENTS

PREFACE

This is the third and final Royal BC Museum Handbook covering
the echinoderms inhabiting British Columbia's shallow coastal
waters. The first, *Sea Cucumbers of British Columbia, Southeast Alaska
and Puget Sound*, appeared 10 years ago. The second book, *Sea Stars
of British Columbia, Southeast Alaska and Puget Sound*, was published
in 2000 (this is a revision of my 1981 book on the sea stars, which
did not cover the adjacent waters in Alaska and Washington).

The first two handbooks described species in two of the five
classes of echinoderms represented in the region. This book covers
the remaining three classes, because together they have only 34
species in this region. I wrote the first two handbooks alone, but
enlisted the help of Dr William Austin for this one. Bill has devoted
more than 40 years to the study of brittle stars and other invertebrates from California to Alaska. His works include a PhD thesis in
1966 on brittle star structure and behaviour, a chapter on brittle
stars in *Intertidal Invertebrates of California*, and a three-volume
annotated checklist of marine invertebrates. Brittle stars can be the
most abundant of readily visible species in some habitats, and their
names are often used to characterize benthic communities. Bill has
studied these communities in Denmark (where they were first
named in the beginning of the last century), and has extrapolated
the results to communities in British Columbia. There is still much
to learn about the dynamics of benthic communities, particularly in
deep water.

As with previous handbooks, we have written this one for stu-
dents, biologists and anyone interested in marine life. The intro-
ductions to each class provide a general background to help

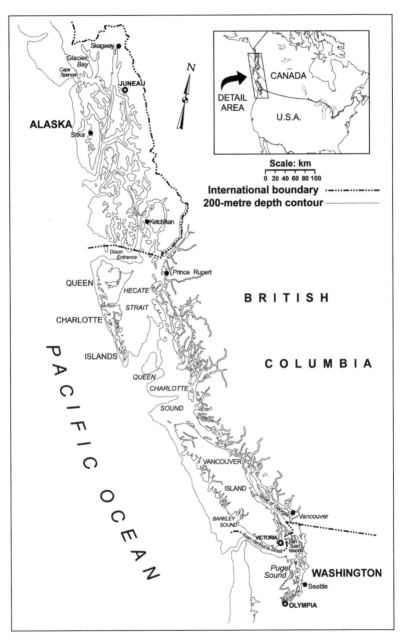

The marine waters of the region covered by this book, showing the 200-metre depth limit.

understand the group. This book covers all the known species of feather stars, sea urchins and brittle stars to a depth of 200 metres from Juan de Fuca Strait and Puget Sound, through British Columbia to Glacier Bay in southeast Alaska. The descriptions are designed so that, with a little effort, anyone can identify an animal to species. To maintain the readability of the text, we have placed references at the end of the species accounts and have listed separately those relating to geographic distribution. At the end of the book, we provide a glossary of technical terms and an extensive reference list of primary sources.

Using all three handbooks, readers should be able to identify every known shallow-water echinoderm in the region.

Philip Lambert

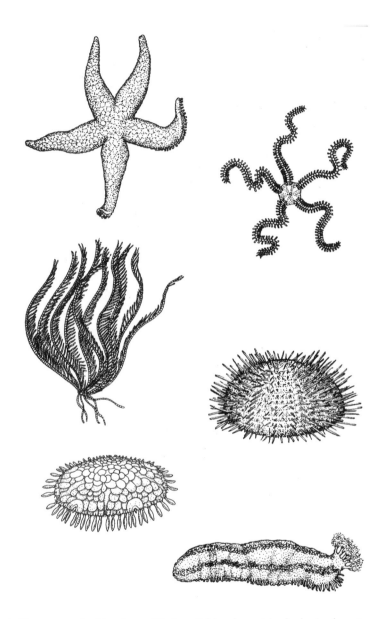

1. Six groups of echinoderms (Phylum Echinodermata), clockwise from top left: sea stars (Asteroidea), brittle stars (Ophiuroidea), sea urchins (Eninoidea), sea cucumbers (Holothuroidea) sea daisies (now considered a subclass of Asteroidea) and feather stars (Crinoidea).

FEATHER STARS (CRINOIDS)

Introduction

Crinoids, known as sea lilies and feather stars, appeared in the fossil record during the Cambrian Period of the Paleozoic Era, 550 million years ago, reaching their peak of abundance in the Mississippian Period, 350 mya. In some places their remains constitute the bulk of the limestone beds laid down during those times. They belong to a group of spiny skinned animals we call Phylum Echinodermata (figure 1). Crinoids can be divided into two basic types: the sea lilies (figure 2), named for their resemblance to floral lilies, attach themselves to the seabed with a stalk; the feather stars (figures 1 and 3), with feather-like arms, have cirri (small prehensile hooks) instead of a stalk for attaching to the seabed. The feather stars are members of the taxonomic Order Comatulida; hence, they can also be called comatulids. Stalked crinoids dominated the seas in the early eras, but their numbers have since diminished, while species of feather stars appear to be on the increase. The majority of about 70 known species of sea lilies exist in deep water below 100 metres, but more than 500 species of feather stars flourish from the tide line to the deep sea.

From Washington to southeast Alaska, we know of two species in one genus between the intertidal zone and 200 metres depth. Five more genera inhabit waters below 200 metres, each with one species; we do not describe them in detail, but include them in the checklist at the beginning of the species accounts.

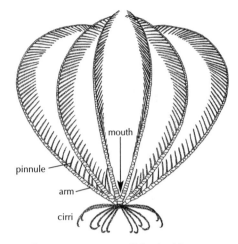

2. Sea lily – a stalked crinoid.

3. Feather star – a comatulid crinoid.

External Anatomy

Unlike most other classes of the Phylum Echinodermata, crinoids usually live with their mouth pointing upwards (figure 3). Most of the body consists of calcareous segments held together by ligaments and covered by a thin layer of living tissue. The central area or crown is covered by soft tissue containing the organs and mouth, and is surrounded by five or more arms. In most species, the mouth is in the centre and the anus is off centre on a small cone. Five ambulacral food grooves radiate from the mouth and onto the arms (figure 4).

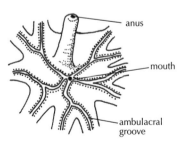

4. The tegmen (mouth region) of a feather star. Podia line the edges of the five food grooves.

Figure 3 shows a generalized feather star. The animal uses the cirri to attach itself to a rocky substrate, but it can release the cirri and change locations. This is the type of crinoid most often found in shallow waters. Deep-water sea lilies attach themselves to the seabed with a long stalk and generally do not move.

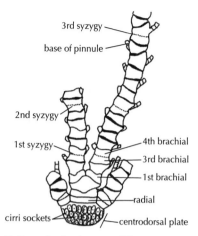

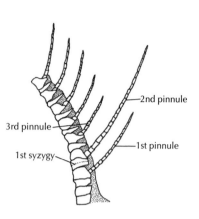

5. Part of a feather star (*Florometra serratissima*) with pinnules and cirri removed showing the basic parts.

6. Positions of the first, second and third pinnules of a feather star (*F. serratissima*).

The brachia (arms), arise around the edge of the oral surface. Primitive forms have 5 arms, but in most species, each arm branches once, yielding a total of 10 arms. Some tropical species have up to 200 arms. The arms are made up of many calcareous pieces, called brachials, which articulate with each other at the joints (figure 5). Some joints are moveable, connected by sets of ligaments and flexor muscles; others are more tightly linked with elastic fibres. One type of immoveable joint is called a syzygy (figures 5 and 6), where two brachials are joined very tightly with ligaments. The opposing faces are ridged, with the ligaments attached in the depressions between the ridges. Externally, the joint appears as a dotted line. The location of this type of joint is important in distinguishing between closely related species. Each brachial has a side branch called a pinnule, with successive pinnules on alternate sides of the arm (figure 5). Often, the last few joints of the pinnule are armed with hooks. An ambulacral groove along the oral surface of the arms and pinnules transports food particles to the mouth (figure 4). Small lobes and clusters of three tube feet situated along the edges of these grooves help to capture water-borne particles and transfer them to the groove.

Internal Anatomy

The central part of the body (the crown) lacks regular muscle layers. Elastic ligaments hold most of the skeletal pieces together. Typical muscle fibres are limited to the arms and pinnules. A pair of flexor muscles that enable the arm to bend inward toward the mouth, connect the brachials. Extension appears to be accomplished by ligaments and flexion by muscles. Recent work indicates that these ligaments have some contractile properties but the mechanism is not known.

The tube feet along the arms and pinnules that capture food particles are part of the water-vascular system, a fluid-filled hydraulic network. Like sea cucumbers but unlike other echinoderms, this system lacks a direct connection to the outside. The water canals from the arms merge into five radial canals under the ambulacral grooves on the tegmen (mouth region, figure 4) that connect to the ring canal surrounding the mouth. From the ring canal, finger-like branches, called labial podia, extend into the mouth opening. On the outer side of the ring canal many short tubes, called stone canals, open into the body cavity. Numerous minute pores, probably serving in respiration, lead into ciliated channels that penetrate the skin around the mouth.

In most crinoids the mouth is central on the disc. It leads directly into a short esophagus, which merges into the intestine. The latter curves in a clockwise direction and returns to the oral surface as the rectum, terminated by the anus on a small papilla. The anal cone of our common species, *Florometra serratissima*, is located off to the side of the tegmen with the mouth in the centre (figure 4).

The crinoid blood system is a network of spaces and sinuses that follow strands of connective tissue running up each arm within branches of the body cavity. Vessels also surround the esophagus with branches to the intestine and down into the centre of the body. There are no blood vessels in the stalk of sea lilies, but branches of the body cavity run down the stalk and carry nutrients and oxygen. Crinoids do not have discrete respiratory organs but simply absorb oxygen through soft tissues of the arms and body, especially the tube feet.

The nervous system consists of three parts that communicate with each other. One has branches running just below the skin along the ambulacral groove. The tracts from all the grooves converge in the tegmen and then run down the wall of the digestive tract. Another network surrounds the mouth outside the water-vascular system and sends ten branches to the arms where they divide in two, one on each side of the arm. These supply the muscles of the water-vascular system, pinnules and tube feet. A central body at the apex of the cavity in the calyx sends out five stout trunks that immediately divide into ten, then into each arm to supply the flexor muscles and the brachials.

There are no apparent sense organs except for the papillae on the tube feet. Free nerve endings in the skin probably serve a sensory function.

Reproduction

Crinoids have separate sexes but there are no obvious differences between males and females. The gonads are located either in the arms or in the genital pinnules near the base of the arms and are simply collections of sex cells in the genital cavity rather than discrete organs. The eggs or sperm escape by rupture of the pinnule wall and they either fall to the bottom or stick to the pinnules with secretions of a cement gland. Some species have brood pouches where the eggs develop into larvae. Males typically spawn first, stimulating the females' discharge of eggs, which are immediately fertilized. Most crinoids breed at a definite season in the year over a period of one or two months, but different species in the same locality may breed at different times.

The fertilized egg develops inside the egg membrane to a uniformly ciliated stage. When the membrane ruptures the larva emerges. Deep-water sea lilies (stalked crinoids) start as a non-feeding auricularia with longitudinal ciliary bands and then develop into a doliolaria larva with cilia around the circumference. Feather stars, on the other hand, skip the auricularia stage, developing directly into the doliolaria. After a few days the uniform cilia reduce to four or five bands encircling the oval embryo. The embryo swims about for a few hours to a few days searching for a suitable substrate. Unlike other echinoderm groups, which have both feeding and non-feeding larvae, crinoids produce only non-feeding larvae even though the eggs are small with very little stored food. The larva attaches to the sea bottom with an adhesive pit at the anterior end, thus the anterior end becomes the stalk or aboral end of the adult. The young crinoid then makes an amazing transformation and the internal organs rotate 90° from the ventral side of the embryo to a posterior, now oral, position. This is the beginning of the pentacrinoid stage – a tiny sea lily with a stalk, only a couple of millimetres in height. The mouth forms and the animal begins to gather food with tiny tube feet around the mouth. Even feather stars, whose adults have no stalk, go through this stage. At six months, the Common Feather Star (*Florometra serratissima*) is about 12 mm tall with an arm span of 6.5 mm and about 250 tube feet. The cirri and pinnules form while the animal is still attached to the stalk, but later, as a juvenile, it detaches from the upper end of the stalk and becomes free-living.

Feeding

Crinoids are passive suspension feeders, meaning that they feed on small particles drifting by in water currents. In deep water, stalked crinoids generally occur in mud where currents are light, but their long stalks elevate the feeding arms off the bottom to where there is some horizontal water flow. Shallow-water crinoids, mostly feather stars, have no stalks but can move to another location to improve feeding opportunities. The details of how they capture food particles have only been studied in a few species. Researchers at the University of Victoria have done much of the work on our local species, the Common Feather Star; see the species account for *Florometra serratissima* for references.

Tube feet situated along the ambulacral food grooves of the pinnules initially collect food particles. The arms and pinnules orient at right angles to the flow with the groove facing down current. Each tube foot is covered with numerous papillae that produce mucus and have sensory hairs at their ends. The tube feet occur in triplets that alternate with those on the opposite side of the groove. The primary (longest) tube feet project out to the side of the pinnule and are exposed to the passing current, the secondary (shorter) tube feet are more vertical and the tertiary tube feet lie across the food groove. The primary tube foot snares food particles with its sticky surface and flicks the particle to the groove, but it will also wipe itself under the tertiary tube foot, which acts like a comb. The particles are moved down the groove by ciliary action and eventually reach the mouth.

Crinoids, where studied, are relatively nonselective in the type of food they eat. Gut contents indicate that they pick up organic and inorganic particles of a similar size. We do know that a large proportion of their food consists of re-suspended bottom detritus containing fecal material, silt and diatom fragments. The nutritional value may come from the microbial populations growing on the organic material.

Parasites and Commensals

Because crinoids are mainly sedentary animals they are subject to infestations by other organisms. Some protozoa, dinoflagellates and ciliates have been found in the digestive tracts. Externally, ciliates have been found gliding among the cirri, hydroids attached to the arms, and brittle stars wrapped around the calyx. Worm-like animals called myzostomes live almost exclusively on crinoids. They are minute, disc-shaped creatures armed with hooks and suckers to cling to the surface. These parasites rob their host of food. Other parasites include carnivorous snails of the family Melanellidae that insert their proboscis and suck out the animal's soft parts. A number of small crustaceans have also been found using a crinoid as shelter and a source of food. These include prawns, isopods, amphipods, copepods and ascothoracids (parasites related to barnacles).

Predators

Fish seem to be the main predators of crinoids in the tropics, where 47 per cent of crinoids had one or more arms missing or regenerating. One researcher reports that 80 per cent of *Florometra serratissima* had one or more arms regenerating; he attributed the damage to predation by the Decorator Crab (*Oregonia gracilis*) and Sunflower Star (*Pycnopodia helianthoides*). Crinoids have evolved a number of strategies to defend against predators. Many tropical species emerge to feed only at night to avoid visual predators, some protect their visceral masses with spine-like oral pinnules or a dense thicket of arms and pinnules, some produce distasteful or toxic compounds, and others swim away. As with many echinoderm relatives, crinoids have great powers of regeneration, allowing them to regrow arms that have been nipped off by fish. In *F. serratissima*, a single arm amputated at the base will regenerate in less than nine months.

Behaviour

Stalked crinoids cannot willfully change their location, but many lose their attachment and move passively in the current. Comatulids have no stalk and can crawl or swim with the aid of their flexible arms. The cirri grasp the substrate then the leading arms pull and the trailing arms push to accomplish crawling. They swim by alternately raising the arms with the pinnules erect then on the down stroke extending the pinnules out to the side. These animals swim for just a few metres, usually because of a disturbance. Many tropical crinoids respond to light by hiding during the day and emerging at night, but no specific light sensor has been identified. Crinoids move their appendages with mutable (contractile) connective tissue (MCT) on the outside of the arm. Only echinoderms are known to possess MCT, which allows them to move without having muscle tissue.

General References

Starfish and Their Relations (Clark 1962) provides a good overview of all the echinoderm groups. The functional anatomy of echinoderms is covered by Lawrence 1987. The main taxonomic reference is a monographic series on existing crinoids by Clark 1915, 1921, 1931, 1941, 1947, 1950, and Clark and Clark 1967. *The Invertebrates* (Hyman 1955) devotes a whole volume to echinoderms. Giese, Pearse and Pearse 1991 reviews each class of echinoderms. Birenheide and Motokawa 1998 describes the unusual contractile connective tissue. Mladenov studied many aspects of the feather star *Florometra serratissima* (Mladenov 1981, 1982, 1983, 1986; Mladenov and Chia 1983). For details of the internal anatomy of crinoids, consult Heinzeller and Welsch 1994.

Checklist of Feather Stars

This checklist contains the names of crinoid species known from southeast Alaska through British Columbia to Washington. The numbers on the right indicate the depth or depth range in metres – the range applies to the species' entire geographic range. Species in bold are described in this book.

Order Hyocrinida
    Family Hyocrinidae
        Ptilocrinus pinnatus A.H. Clark, 1907   2904

Order Bourgueticrinida
    Family Bathycrinidae
        Bathycrinus pacificus A.H. Clark, 1907   1655

Order Comatulida
    Family Pentametrocrinidae
        Pentametrocrinus cf. varians (P.H. Carpenter, 1888)   660–1302
    Family Antedonidae
        Florometra asperrima (A.H. Clark, 1907)   79–1574
        Florometra serratissima (A.H. Clark, 1907)   11–1252
        Psathyrometra fragilis (A.H. Clark, 1907)   439–2903
        Retiometra alascana A.H. Clark, 1936   291–1270

Key to Crinoids of British Columbia, Southeast Alaska and Puget Sound

1a. Crinoids with a stalk attaching it to the substrate (figure 2). 2
1b. Unstalked crinoids with cirri at base (figure 3). 3

2a. Five arms. . . . .*Ptilocrinus pinnatus*
2b Ten arms. . . . .*Bathycrinus pacificus*

3a. Five arms. . . . .*Pentametrocrinus* cf. *varians*
3b. Ten arms. 4

4a. Second pinnule slightly longer than the first (figure 6). 5
4b. Second pinnule shorter than the first. .6

5a. Third syzygy (figure 5) at brachials 14-15.
. . . .*Florometra asperrima* (page 15)
5b. Third syzygy at brachials 16-17.
. . . .*Florometra serratissima* (page 16)

6a. Arms up to 130 mm long; three or four columns of cirri sockets per radial; 25–35 cirral segments (figure 7).
. . . .*Psathyrometra fragilis*
6b. Arms up to 75 mm long; 11–12 cirral segments.
. . . .*Retiometra alascana*

Species Accounts

Family Antedonidae

The largest family of existing feather-star crinoids, mostly ten-armed species. Slender, delicate outer pinnules; wedge-shaped brachials; regularly spaced syzygies.

Florometra asperrima

Description
Arms: Ten, up to 235 mm long with 330 joints or brachials; first pinnule, 31 mm long with 63 joints, slender, tapering gradually to a delicate tip; second pinnule, 32–33 mm with 65 joints; third pinnule 33 mm with 55 joints. Joint edges are raised distally, overlapping the succeeding joint. Distal joint edges set with a row of fine sharp teeth. Syzygies between brachials 3-4, 9-10, 14-15, and after that every three muscular articulations.
Cirri: 42–53 moderately stout, 45–70 mm long, the first segment is about 2.5 times as broad as its length.
Colour: Yellow or tan; cirri whitish.
Taxonomic Notes: Although originally described as a species distinct from *Florometra serratissima* some taxonomists feel that the differences between these two species are not sufficient to separate them, because the range in position of the third syzygy in the two species overlaps each other. The name *asperrima* is from the Latin *asper*, meaning "rough, harsh".

Similar Species

Almost identical to *Florometra serratissima*. The position of the third syzygy is the main diagnostic difference. *F. asperrima* has a third syzygy between brachials 14-15, whereas *F. serratissima* has it between brachials 16-17.

Distribution

Aleutian Islands in the southern Bering Sea to Monterey Bay, California, and from the Okhotsk Sea to Tsugaru Strait, Japan. In deep water in the south, shallower in the north, 79–1574 metres. Records from Skidegate Channel, Sumner Strait (southeast Alaska) and off Triangle Island, but primarily an Aleutian species.

Biology

Not known.

References

Clark 1907, 1937; Clark and Clark 1967.

Florometra serratissima   Common Feather Star

Description

The most common species in British Columbia and southeast Alaska, possessing the basic comatulid body plan as described in the introduction (pages 6–7).

Arms: Ten, 150–280 mm long; first pinnule 17–21 mm long with 45–60 joints; second pinnule longer than the first, 18–22 mm with 45–60 joints; third pinnule 20 mm with 36 joints; distal pinnules long and slender, their joints overlapping. Syzygies between brachials 3-4, 9-10, 16-17, and after that every three muscular articulations (figure 5). Each arm has approximately 440 pinnules.

Cirri: 40–50 large and stout, about 30 mm long with about 36 joints; nearly all the cirrals bear dorsal spines.

Colour: Overall yellow, brownish-yellow, tan or reddish-brown; cirri whitish to light brown; pinnules brown to black. **See colour photos C-2 and C-3.**

Taxonomic Note: The name *serratissima* is from the Latin *serratus*, meaning "saw teeth".

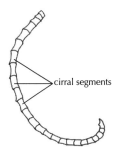
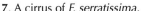
cirral segments

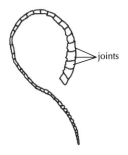
joints

7. A cirrus of *F. serratissima*. **8**. A pinnule of *F. serratissima*.

Similar Species

Florometra asperrima is almost identical in general appearance to *F. serratissima* except it tends to be more yellow in colour, larger and the third syzygy is between brachials 14-15 instead of 16-17. Charles Messing thinks that these two species may be the same.

Distribution

Shumagin and Sannak islands, Alaska, to Natividad Island, Baja California: 11–1252 metres.

Biology

Feeding: *Florometra serratissima* eats particulate matter either drifting up from the sea bottom or suspended in the current. Tube feet situated along the ambulacral food grooves of the pinnules initially collect food particles. The arms and pinnules orient at right angles to the flow of water, with the groove facing down current. Each tube foot is covered with numerous papillae that produce mucus and have sensory hair-like cilia at their tips. The tube feet occur in triplets that alternate with those on the opposite side of the groove. The primary (longest) tube feet project out to the side of the pinnule and are exposed to the passing current, the shorter secondary tube feet are more vertical and the tertiary tube feet lie across the food groove. The primary tube foot snares food particles with its sticky surface and flicks the particle to the groove but it will also wipe itself under the curled tertiary tube foot, whose papillae act like a comb. The particles are moved down the groove by ciliary action and eventually reach the mouth.

Reproduction: The sexes are separate and gonads occupy the pinnules on the lower part of the arms. The development of the gametes appears to be synchronized throughout the animal and the release of eggs and sperm is more-or-less continuous through the year. The mature adults release gametes through nipple like swellings on the distal surface of each genital pinnule. Individual females ovulate about 23,800 eggs for a period of about three days each month. The eggs of this and other crinoids are small (around 200 μm) compared to other echinoderms that have non-feeding larvae. Sperm fertilize the eggs in open water and development begins with radial cleavage.

Thirty-five hours after fertilization the young larva (doliolaria) breaks through the fertilization membrane surrounding the egg. By four days, the larva has transverse bands of hair-like cilia for swimming and an adhesive pit at the front end for attaching to the substrate. Soon after this, the larva begins to explore the sea bottom for a place to settle, which could happen right away or be delayed by up to nine days. In artificial culture larvae tend to settle next to other attached individuals. This likely occurs in the wild, also, resulting in large aggregations of adults.

Once settled and attached to the substrate, the larva undergoes some rapid changes and resembles a small stalked crinoid (the cystidean stage), although it does not have a mouth yet. About 16 days after settlement, when the larva is about 1.8 mm tall, another transformation takes place. Plates and tube feet begin to form in the mouth region and the animal (in its pentacrinoid stage) begins to feed. It survives in this form for at least six months. Although the transformation into a free-living form has not been observed, in other species of crinoid the hook-like cirri develop around the top of the stalk, then the animal detaches from the stalk and leaves it behind.

Other Notes: At least six species of sea stars and two sea anemones cause *Florometra serratissima* to swim for short bursts of 10 to 30 seconds. Locomotion is by a sequential repetition of arm strokes in three groups of arms. The initial response raises the animal about 30 cm off the substrate, then it swims horizontally with mouth foremost at a mean speed of 6.8 cm per second. Feather stars will also crawl to improve their feeding position. The leading arms pull while the trailing arms push the body to a new position usually higher on a rock where the current is better.

As with other species of echinoderms, this species can regenerate parts of the body that have been damaged. A single arm amputated at the base regenerated to a functional one within nine months. The Sunflower Star (*Pycnopodia helianthoides*) and the Decorator Crab (*Oregonia gracilis*) prey on this crinoid, and are thought to cause it to cast off its arms as a defense.

The parasitic flatworm *Fallocohospes inchoatus* inhabits the intestine of *F. serratissima*.

References
Byrne and Fontaine 1981, 1983; Clark 1907; Clark and Clark 1967; Hendler 1996a; McDaniel 1976; McEdward, Carson and Chia 1988; Messing and White 2001; Mladenov 1981, 1982, 1983, 1986; Mladenov and Chia 1983; Neve and Howard 1970; Scouras and Smith 2001; Shaw and Fontaine 1990; Shinn 1986; Webster 1975. **Range:** Clark and Clark 1967.

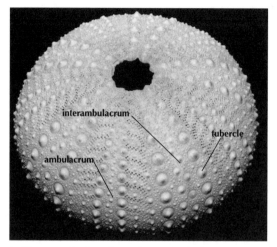

9. The test of a Purple Urchin (apical plates removed).

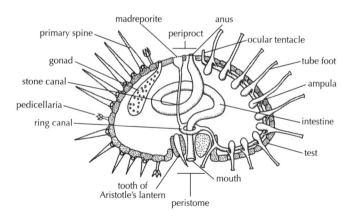

10. The anatomy of a sea urchin.

SEA URCHINS (ECHINOIDS)

Introduction

The fossil record shows that sea urchins first appeared in the middle Ordovician Period about 500 million years ago, although their relatives appeared earlier. By the late Ordovician there were already 765 genera. Since the Triassic (230 mya) this class has evolved rapidly, with many new and distinctive shapes.

At present, between Puget Sound and southeast Alaska, we know of eight species in four genera between the intertidal and 200 metres depth. Another eleven species in nine genera occur only below 200 metres – we show them in the checklist but do not describe them in the species accounts.

External Anatomy

Based on their shape, the echinoids are informally divided into two categories. The regular sea urchins are radially symmetrical, either globular or ellipsoid (figures 1 and 9), with the anus located within the plates of the apical system at the top and the mouth usually on the underside (figure 10). The irregular sea urchins (figures 23 and 24) are usually elongated or disc-like and bilaterally symmetrical with the anus outside the apical system. The Green Sea Urchin (colour photograph C-6) is an example of a regular urchin, and the Heart Urchin (colour photograph C-11) an irregular urchin.

The test (shell) of a regular sea urchin is made up of a series of plates interlocking into a solid almost spherical skeleton (figure 9). Other species have evolved different shapes depending on their habitat. For example, a sand dollar has a disc-shaped test, a heart

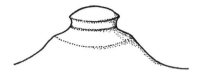

11. A mamelon that supports a primary spine of a Pink Sea Urchin (*Strongylocentrotus fragilis*).

urchin has an elongated or egg-shaped test, and a deep-sea soft urchin has a poorly calcified test that collapses when out of the water. Each plate in a sea urchin's test has a crystalline core to which layers are added as the animal grows. A microscopic view shows concentric lines of growth on each plate. The plates run in 20 vertical rows from the mouth to the anus, like lines of longitude on a globe, two ambulacral rows (figure 9) with sets of pores for tube feet alternating with two interambulacral rows lacking pores (figure 9). The number of pores is an important taxonomic character for separating species (see figures 17, 19 and 21). In regular urchins each pair of pores gives rise to one tube foot, but in irregulars, tube feet arise from single pores.

Most plates bear a mamelon (a knob-like tubercle), which articulates with a socket at the base of a spine. Mamelons may be rounded and plug-like or be undercut like the cap of a mushroom (figure 11). The spines are attached and moved by two layers of radiating muscles (figure 12). The muscles move the spine and an inner connective tissue layer holds the spine rigid. This mutable collagenous tissue is peculiar to echinoderms and is capable of becoming stiff under nervous control. In cross-section, spines are made up of a series of wedges; the number of wedges is characteristic for a species (see figure 18).

Pincer-like appendages, called pedicellariae (figure 13), also attach to small tubercles on the test. The pedicellariae react when something foreign touches the surface of the urchin. They can be aggressive or defensive (see Feeding and Predators) or simply keep the surface clean by crushing particles and allowing the surface cilia to sweep the bits away.

At the top end of each ambulacral row is an ocular plate, which contains a single pore for

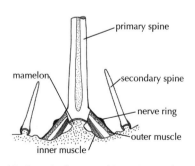

12. A typical sea-urchin spine and mamelon.

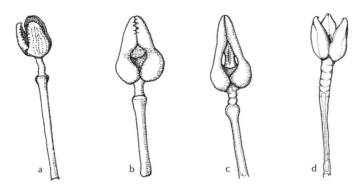

13. Three types of pedicellariae: **a**, ophicephalous of *Strongylocentrotus droebachiensis*; **b**, tridentate of *S. droebachiensis*; **c**, tridentate of *S. franciscanus*; and **d**, globiferous of *S. droebachiensis*.

an ocular tentacle (figure 14). The name implies a visual function but there is no evidence that this tentacle responds to light. Still, some urchins react to shadows of predators, so some part of the epidermis is obviously sensitive to light. Genital plates at the top end of the interambulacral columns alternate with the ocular plates. Each one has a gonopore through which eggs and sperm are extruded during spawning (figure 14). One of these plates is modified into the madreporite (figures 14 and 15). This circle of ocular and genital plates constitutes the apical system. Inside this circle a cluster of small plates surrounds the anus, together referred to as the periproct (from the Greek *peri*, meaning "near" or "around", and *proktos*, referring to the anus or rectum).

The other end of the test, where the mouth opens, is called the peristome (after the Greek word *stoma*, meaning "mouth"). The

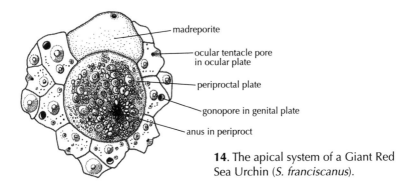

madreporite

ocular tentacle pore in ocular plate

periproctal plate

gonopore in genital plate

anus in periproct

14. The apical system of a Giant Red Sea Urchin (*S. franciscanus*).

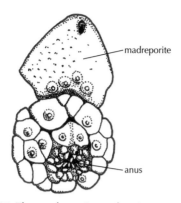

15. The madreporite and periproct of a Green Sea Urchin (*Strongylocentrotus droebachiensis*).

flexible buccal membrane stretches across the opening in the test and the mouth is in the centre. Some urchins have small plates embedded in this structure, but in most the membrane is flexible so that it can be expanded and contracted to allow the jaws to move up and down. The buccal membrane is moved by a set of circular and radiating muscles and the muscles of the jaw apparatus.

The outer epidermis, a single layer of ciliated cells equipped with tiny moveable hairs (cilia) covers all external parts including all the appendages (spines and pedicellariae). A nerve network is situated beneath the epidermis but outside the test. Each spine and pedicellaria has a nerve ring around the base with branches travelling up the length.

Internal Anatomy

Most regular urchins have a mouth apparatus shaped like an old-fashioned lantern (or like an inverted five-sided pyramid). It has five teeth operated with a complex set of muscles and levers, 40 skeletal pieces in all. The Greek philosopher Aristotle was the first to describe the sea urchin's mouth, so it is called Aristotle's lantern. (For a more detailed description of Aristotle's lantern, consult: Hyman 1955; Durham et al. 1966; Kier 1987; and Brusca and Brusca 1990.) This organ varies in size and complexity and in some groups, is completely absent.

Like all echinoderms, echinoids possess a hydraulic system for operating a series of tube feet (podia) that protrude from the five sets of pore pairs running from the oral to aboral poles of the test (figure 10). This water-vascular system consists of a ring canal encircling the intestine just above the Aristotle's lantern with branches running down the inner surface of the lantern, then following the inside curve of the test providing branches through the test to each tube foot. Just inside the test, each foot has a valve and a balloon-like ampulla that controls the length of the tube. It then passes through the test via a pair of pores and reunites into one tube foot on the outside. In regular urchins each pore pair represents one tube foot. The ring canal also gives off a main branch, called the stone canal, that connects to a special perforated plate, the madreporite (figure 15), at the aboral pole of the test (figure 10). As in the sea stars, this is thought to top up the fluid in the system. The flexing of the soft membrane around the mouth probably compensates for fluctuations in volume in other parts of the system.

The digestive tract is the most obvious organ in the interior of the test. From the mouth, it passes through the lantern as the pharynx. It emerges from the top as the esophagus and forms a short loop. Then it becomes the small intestine, which when viewed from the top spirals anticlockwise. After almost forming a complete circle it doubles back on itself and expands into the large intestine, which winds around the space, exiting at the anus. The intestine is suspended from the inside of the test by thin, membranous mesenteries.

At certain times of the year the ripe gonads are a significant part of the internal anatomy of a sea urchin. The female ovaries or male testes occupy the inner surface of the interambulacra, running from the lantern to the apex of the test. At the aboral end, short gonoducts pass to the outside via the gonopores in the genital plates of the apical system.

The spaces in the body are filled with fluid similar to seawater and contain amoeboid cells that also wander through the body tissues. Echinoids have six types of blood cells each with a specific function – some attack foreign intruders, others produce pigment. They also have a haemal (blood) system, but the vessels are small, and primarily supply the gonads with nutrients from the gut. Branches of the haemal system also extend along the water and nerve rings and most of the major organs.

Nervous System

The echinoid nervous system consists of a circumoral nerve ring, radial nerves following the ambulacral rows, and a network of nerves under the skin. The circumoral nerve ring supplies the mouth region, with nerves ascending the digestive tract and other branches to the inner surface of the body wall along the midline of each ambulacrum. The radial nerves branch off to each tube foot, then ascend it to spread out a nervous layer on the terminal sucker. The tube feet can be extended a great deal to act like feelers or antennae as the animal advances along the sea floor. Nerve branches also pass through the podial pores and spread out in the body wall on the outside of the plates in a broad network, including the spines and pedicellariae. Much of the body surface, including the spines and pedicellariae, is richly supplied with sensory nerves. In 1872, Sven Lovén discovered and described small organs, called sphaeridia, that are almost invisible to the naked eye; they appear to be modified spines, oval in shape and attached to a small stalk. They occur in the midline of the ambulacral rows on the oral side of the test. Most researchers believe they are organs of equilibrium.

Reproduction and Development

Regular echinoids have five gonads suspended by mesenterial strands along each of the five interambulacra (figure 9). At the aboral end, a gonoduct exits by the gonopore situated on the genital plate of the apical system (figure 14). Irregular echinoids, such as the Heart Urchin, usually have only four gonopores and four gonads. Echinoids have separate sexes, with some rare exceptions.

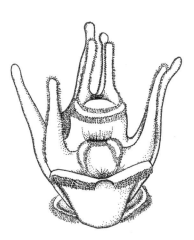

It is not possible to tell the sex from external characters except in some brooding species or perhaps by examining the size of the gonopore papillae. Ripe individuals shed eggs and sperm directly into the seawater where fertilization takes place. In temperate regions most spawning takes place in the spring and summer, and some sea urchins brood their young.

Upon fertilization, cleavage takes place fairly rapidly to the eight-cell stage. At the hollow blastula stage the cells produce whip-like cilia causing the embryo to rotate within the fer-

16. The echinopluteus larva of a Giant Red Sea Urchin (*Strongylocentrotus droebachiensis*).

tilization membrane, which ruptures and releases the embryo. A typical embryo develops into a larval type called an echinopluteus, with long arms supported by rods of calcite (figure 16). Bands of cilia develop around the bases of the arms and allow the embryo to swim up and down in the water and gather food. In four to six weeks, depending on the species, the water temperature and the availability of food, the embryo undergoes metamorphosis to a young urchin. The covering of the larval arms is absorbed and the skeletal rods may be absorbed. The mouth and anus of the larva close over. The juvenile urchin, about 1 mm in diameter, moves by means of the first five pairs of tube feet, and a number of other complex changes take place internally as the larva transforms its body from bilaterally symmetrical to radially symmetrical. The adult mouth forms and larval organs are resorbed and reorganized as the animal begins to feed and grow.

Feeding

Sea urchins eat seaweeds and animals. They use their sense of smell to locate food, detecting chemicals given off by their prey. Most of the animals they eat are attached to the substrate, like sea squirts, hydroids and sponges. In some, the pedicellariae can paralyze small animals and larvae that try to attach to the outer surface, or grab the hairs of some crustaceans, allowing the urchin to capture live food. On our coast, sea urchins tend to eat seaweeds. In the fall we often see large kelps covered with sea urchins munching away on their tissues. They use their spines and tube feet to grab pieces of the algae then manoeuvre them down to the jaws that protrude through the mouth at regular intervals to tear off smaller pieces. Some heart urchins live in a burrow in the mud with an open shaft to the surface. Special tube feet extend up to the surface and search the mud for particles of food, which adhere to mucus. Other heart urchins burrow slowly through the mud and the tube feet deliver food particles to the mouth. Most sand dollars pick up sand grains or food particles with their tube feet and pass them to the mouth via the food grooves.

Behaviour

Most sea urchins retreat into the shade during the day. Many species attach stones and pieces of shell to their aboral surface. One explanation is that the items protect them from strong light, but it may also minimize water loss and conceal them from visual predators. A sudden change in the light causes many urchins to move their spines to the erect position in an apparent defensive reaction. The tropical genus *Diadema* has very active spines that move back and forth when the animal is disturbed. Lambert once made the mistake of trying to pick one up by a spine to pull it out of a tide pool and it speared the end of a finger – it stung for about 30 minutes! Spines of the local Giant Red Sea Urchin (*Strongylocentrotus franciscanus*) can penetrate the neoprene suits of divers. They may break off in the skin and fester for weeks. The tropical genus *Toxopneustes* has poison-containing pedicellaria and contact can be fatal to humans.

Parasites and Commensals

Many small animals use the spine canopy of urchins for shelter. Some, such as tubeworms, attach to the spines of pencil urchins, while scale worms, ostracods, shrimp and small crabs roam on the outer surface among the spines and pedicellariae. Tiny bivalves and amphipods may be associated with burrowing heart urchins and some parasitic snails bore into the test of urchins and deposit eggs or feed on gonadal tissues. A parasitic barnacle consisting primarily of a reproductive sac that opens through a pore in the test, releases its larvae to the exterior. Parasitic copepods use their mouthparts to pierce the skin and suck out bodily fluids. Internally, the intestine harbours unicellular ciliates, flatworms, nematodes and trematodes; most cannot survive outside of the host.

Predators

Despite their protective spines, some predators routinely prey on sea urchins. Vertebrates that eat sea urchins include Sea Otters (*Enhydra lutra*), crows, gulls, Wolf-eels (*Anarhichthys ocellatus*) and other large fish. Sea stars and crabs are the main invertebrate predators. Sea urchins use their spines and pedicellariae to defend themselves. Globiferous pedicellariae (figure 13d) have poisonous glands that will repel the tube feet of an attacking sea star. Usually the pedicellariae will break off as the urchin moves away from its attacker. Tridentate (three-jawed) pedicellariae (figure 13b, c) will bite any objects that come in contact with them. Others keep the surface clean by crushing particles and allowing the surface cilia to sweep the bits away.

Of all the predators, humans are probably the most devastating. People in many parts of the world, such as the Mediterranean, Peru, Ecuador and Japan, consider the sea urchin's ripe gonads a delicacy. On this coast, fishers harvest the Giant Red Sea Urchin and ship the fresh product to Southeast Asia and elsewhere in North America.

General References

For a good overview of all the echinoderm groups consult *Starfishes and Related Echinoderms* (Clark 1977). Several popular books that describe common echinoderms of the west coast of North America include *Between Pacific Tides* (Ricketts et al. 1985), *Intertidal Invertebrates of California* (Morris et al. 1980) and *Seashore Life of the Northern Pacific Coast* (Kozloff 1983). Hyman 1955 is a classic that is hard to beat for details on all aspects of echinoids and other echinoderms, although it may be a bit dated on some details.

For more technical aspects of echinoids consult the following: Mortensen 1928 (and later volumes up to 1951) is the most important taxonomic reference; D'yakonov 1969 is an English translation of his earlier 1950 publication in Russian, which includes 85 pages of detailed introduction to echinoids; Moore edited the series *Treatise on Invertebrate Paleontology* (1966), which includes Parts S, T and U on echinoderms; *Physiology of Echinodermata* (Boolootian 1966) is a classic text on the group; Strathmann 1987 describes the reproduction of echinoids found off the coast of northwestern North America, and Giese et al. 1991 covers reproduction in echinoderms worldwide; Cavey and Markel 1994 reviews the microscopic anatomy of echinoids. Information in these references has been updated in individual scientific papers published subsequently. Each species description lists specific papers with the most recent information.

Checklist of Sea Urchins

This checklist contains the names of echinoid species known from southeast Alaska through British Columbia to Washington. The numbers on the right indicate the depth or depth range in metres – the range applies to the species' entire geographic range. Species in bold are described in this book.

Order Cidaroida
    Family Cidaridae
        Aporocidaris milleri (A. Agassiz, 1898)   300–3937

Order Echinothurioida
    Family Echinothuriidae
        Sperosoma biseriatum Doderlein, 1901   1864–3230
        Sperosoma giganteum Agassiz & Clark, 1907   1211

Order Echinoida
    Family Stronglylocentrotidae
        Strongylocentrotus droebachiensis (Müller, 1776)   0–300
        Strongylocentrotus fragilis Jackson, 1912   50–1260
        Strongylocentrotus franciscanus (Agassiz, 1863)   0–125
        Strongylocentrotus pallidus (Sars, 1871)   5–1600
        Strongylocentrotus purpuratus (Stimpson, 1857)   0–161

Order Clypeasteroida
Suborder Scutellina
    Family Dendrasteridae
        Dendraster excentricus (Eschscholtz, 1831)   0–90

Family Echinarachniidae
Echinarachnius parma (Lamarck, 1816) 0–150

Order Holasteroida
Family Urechinidae
Antrechinus drygalskii perfidus (Mironov, 1976) 4990–5740
Cystechinus loveni (A. Agassiz, 1898) 1571–4800
Family Pourtalesiidae
Ceratophysa valvaecristata Mironov, 1975 4200–6320
Cystocrepis (Echinocrepis) setigera (Agassiz, 1898) 2876–4072
Echinocrepis rostrata Mironov, 1973 3315–5020
Pourtalesia tanneri Agassiz, 1898 1450–3954
Pourtalesia thomsoni Mironov, 1975 3315–4321

Order Spatangoida
Family Schizasteridae
Brisaster latifrons (A. Agassiz, 1898) 20–1800

Family Aeropsidae
Aeropsis fulva (A. Agassiz, 1898) 1455–5390

Selected References: Mironov 1976; Mooi and David 1996; Mortensen 1928, 1948, 1951.

Key to Echinoids of British Columbia, Southeast Alaska and Puget Sound

1a. Body round and flat, like a cookie, covered with short, fuzzy spines. 2
1b. Body not flat. 3

2a. Petal-like pattern in the centre of the dorsal surface; food grooves on ventral side, long and straight, only branching near the perimeter. . . . .*Echinarachnius parma* (page 51)
2b. Petal-like pattern off-centre with posterior leaves shorter; food grooves extensive and branching and also on the dorsal side of the test. . . . .*Dendraster excentricus* (page 48)

3a. Spine-covered, egg-shaped body, with mouth at one end; usually living in mud. . . . .*Brisaster latifrons* (page 54)
3b. Spines covering a nearly spherical test; mouth on the underside. 4

4a. Primary spines attach to tubercles with a straight-sided mamelon, like a plug. 5
4b. Primary tubercles have a mamelon with an undercut neck. .6

5a. Primary spines usually have 50–100 wedges; eight or nine pore pairs to an arc; large body with long, sharp spines, reddish in colour. . . .*Strongylocentrotus franciscanus* (page 40)
5b. Primary spines have 20–55 wedges; eight pore pairs to an arc; medium-sized body with spines that are moderate in length, stout and blunt; generally purple. . . . .*S. purpuratus* (page 45)

6a. Primary spines composed of 30–35 wedges; five pore pairs to an arc; generally pinkish. . . . .*S. fragilis* (page 37)
6b. Primary spines have fewer than 35 (18–27) wedges. 7

7a. Usually six or seven pore pairs per arc; generally white, with white tube feet. . . . .*S. pallidus* (page 43)
7b. Usually five or six pore pairs; greenish with dark tube feet. . . . .*S. droebachiensis* (page 34)

Species Accounts

Family Strongylocentrotidae

Regular sea urchins of moderate to large size. Test has no depressions or pits; mamelons are smooth; more than three pore pairs to an arc; globiferous pedicellariae have no lateral teeth.

Strongylocentrotus droebachiensis   Green Sea Urchin

Description

Strongylocentrotus droebachiensis is a medium-sized urchin about 80 mm in diameter with a low but not flat test, and medium-length spines about one-fifth the diameter of the test.

Colour: Generally greenish-brown. Light green at base of the spines and lighter toward the tip. Vertical rows of tube feet usually appear as dark bands. The test is usually purple to violet rather than green. **See colour photo C-6.**

Pore pairs: Usually five in each arc, but occasionally six.

Spines: Primary spines are not conspicuously larger than secondary spines and form uniform vertical rows. The mamelon has an undercut neck. The number of wedges of the primary spines seems to vary geographically, but most urchins up to a diameter of 65 mm have 23–27 wedges, some up to 33.

Pedicellariae: Globiferous have a muscular neck and ovoid head (figure 13d); tridentates have narrow-leaved blades with parallel sides, a few saw teeth at the tip of each and the head forming a tapered conical shape. The ophicephalous type are similar throughout the genus, so not diagnostic.

Taxonomic Notes: On the Pacific coast of North America the Green Sea Urchin was first described as *Echinus chlorocentrotus* by Brandt, 1835, from Sitka, Alaska. Mortensen 1910 found that the urchin was not physically different from *Strongylocentrotus droebachiensis* and concluded that *E. chlorocentrotus* should be reduced to a junior synonym. Biermann 1998 found that the DNA of *S. droebachiensis* from Europe (northeast Atlantic) differed by three per cent from that in North America (the western Atlantic and the northeast Pacific). The name *droebachiensis* is after the town of Drøbak, Norway, where the species was first described.

Similar Species

In this region, *Strongylocentrotus pallidus* could be confused with *S. droebachiensis*. Generally, colour will separate these two species – *S. pallidus* is white and *S. droebachiensis* is greenish – but if the colour differences are not obvious, check the number of pore pairs and the shape of the globiferous pedicellariae: *S. pallidus* has six or seven pore pairs and spherical-headed globiferous pedicellariae; *S. droebachiensis* has five pore pairs and ovoid globiferous pedicellariae.

Distribution

A widespread Arctic and northern boreal species; from the Arctic Ocean to Washington and the Sea of Japan in the Pacific, and from Hudson Bay, Greenland, Iceland, northern Europe to Chesapeake Bay, USA, Scotland and the western part of Baltic Sea in the Atlantic. 0–300 metres, but most common in 0–50 metres. RBCM collection 0–155 metres.

Biology

Feeding: Feeds primarily on fixed algae and depends on season and locality, but also on small gastropods, barnacles, dead fish, diatoms and detritus. Feeding rates vary with different species of algae. For example, an urchin can eat Bull Kelp (*Nereocytstis luetkeana*) at a rate of 207 mg/hour, but thin green Sea Lettuce at only 17 mg/hour.

Reproduction: In the San Juan Islands individuals can be ripe from January to June with a peak in spawning from March to April. Eggs are usually 155–160 μm in diameter. At 9–10°C the embryo reaches the pluteus stage in five days and metamorphoses into a juvenile urchin in 9 or 10 weeks. The test reaches a diameter of one centimetre in about a year. All these values can vary depending on temperature and food supply.

Other Notes: Two species of flatworms, *Syndesmis inconspicua* and *Syndisyrinx franciscanus,* inhabit the intestines of *Strongylocentrotus droebachiensis.* In Norway this species is infected with the nematode *Echinomermella matsi.* This parasite has only a minor effect on the urchin's growth rate, but it reduces its host's life expectancy by 33–86 per cent. Experiments on the east coast of Canada showed that aggregations of this sea urchin were primarily the result of attraction to a common food such as algae rather than a defensive reaction to predators such as crabs or lobsters; in the absence of food, individuals would flee from certain predators.

Based on mitochondrial DNA analysis, some researchers conclude that *S. droebachiensis* colonized the Atlantic from the Pacific Ocean – 80 per cent of individuals are in one of two genotypes. Green Sea Urchins in the eastern Pacific and western Atlantic differ in mitochondrial DNA by only 0.1 per cent, but between the western Atlantic and the eastern Atlantic (Norway and Iceland), the divergence is about 3 per cent, suggesting they have been isolated from each other for a much longer time. Studies also suggest that western Atlantic sea urchins are descendants of a rare migratory group that colonized from the Pacific within the past 300,000 years, but the first entry was thought to be three to five million years ago.

Fisheries

The roe of Green Sea Urchins has long been a delicacy, particularly in Japan. These urchins were commercially harvested sporadically along the west and east coasts of Canada and the northern United States from the 1950s into the 1980s. In Maine, harvesting followed a typical boom-and-bust cycle: from 2400 tons in 1988 up to 16,000 in 1993, then dropping back to 2400 tons by 2002 when the harvest levelled off. Similar increasing harvests occurred in Nova Scotia and New Brunswick with a levelling off or decline by 1995.

The urchin fishery in British Columbia harvested about 250 tons in 1986, peaked at about 1000 in 1992, then dropped back to 250 by 1995. The harvest was capped at about 175 tons per year in 1996. A small fishery also occurs near Kodiak, Alaska.

The potential overharvesting of wild stock has generated pilot studies on the aquaculture of Green Sea Urchins over the past few years.

Urchin Barrens
The Green Sea Urchin periodically undergoes population explosions in British Columbia and Nova Scotia. A dense population of urchins can graze off all the kelps and most of the other non-calcareous seaweeds on the bottom. Urchins will also eat much of the animal life attached to the bottom. This overgrazing transforms a kelp forest into an open, level bottom with low species diversity. Some biologists have proposed that decreases in predators, such as the Sea Otter (*Enhydra lutra*), have caused population explosions, but others disagree. Another possibility is a periodic increase in water temperature, resulting in more rapid growth and settlement of larvae with associated decreases in larval predation.

References
de Ridder and Lawrence 1982; Jensen 1974; Palumbi and Wilson 1990; Strathmann 1987; Stien 1999; Westervelt and Kozloff 1992; Vadas, Elner, Garwood and Babb 1986. **Range:** Jensen 1974; Bazhin 1998.

Strongylocentrotus fragilis   Pink Sea Urchin

Description
Strongylocentrotus fragilis has a thin, fragile test up to 80 mm in diameter with fine, needle-like spines about one-fifth of the test diameter.
Colour: Pinkish-orange. The spines are orange at the base grading into white at the tip. The test is rose red or pink with a violet cast on the upper side and more violet orally. Pedicellariae are white. Tube feet are pale with two dark lines running lengthwise. **See colour photo C-8.**
Pore pairs: In arcs of 5 or 6.
Spines: Primaries have 30 to 35 wedges in cross-section; base of spine sits on a mamelon with an undercut neck (figure 11).
Pedicellariae: Three-jawed with slender valves, a long blade and a round base.
Taxonomic Notes: Originally described as *Strongylocentrotus fragilis* Jackson; revised to *Allocentrotus fragilis* by Mortensen (1942), but Biermann (1998) and Biermann et al. (2003) have shown with DNA

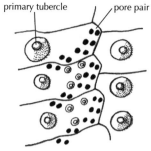
primary tubercle   pore pair

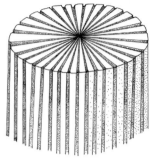

17. The arrangement of pore pairs in *S. fragilis.*

18. Cross-section of a primary spine of *S. fragilis*, showing the wedges.

analysis that it is a close relative of *S. pallidus* and should be a member of the genus *Strongylocentrotus*. The Latin *fragilis* means "easily broken".

Similar Species
Resembles *Strongylocentrotus pallidus* and *S. droebachiensis* in general shape, but they can be distinguished by their colour, thickness of spines and number of wedges in the spines.

Distribution
West coast of North America from the Queen Charlotte Islands to Baja California; 50–1260 metres. Most specimens in Royal BC Museum's collection are from the outer continental shelf (west of the main coastal islands) with only two from inlets. Usually dredged from soft bottoms, but we have also observed them with an ROV on rocks and cobble and mixed with *Strongylocentrotus pallidus*. RBCM collection 115–550 metres.

Biology
Feeding: Bottom detritus such as decomposing seaweeds, diatoms, sponge spicules and foraminiferans.

Reproduction: Off Monterey, the gonads of this species are well developed from September until they spawn in January or February; but the state of the gonads varies considerably in the population. In another study, off Oregon, data suggests that *S. fragilis* had a semiannual spawning in early spring and early autumn.

Below 400 metres no reproductively mature individuals were collected. At Friday Harbor, San Juan Island, this species took 141–252 days after fertilization to settle in artificial culture.

Other Notes: Growth rates appeared to be similar between 100 and 600 metres depth, but decreased between 800 and 1260 metres. Based on size-frequency distributions of populations and lines of growth on the calcareous plates, researchers have calculated the maximum age of this species at 7.5 years. This sea urchin is attracted to light, but seems to prefer low-intensity light. The greatest sustained speed when stimulated by light was 3.3 metres/hour (5.5 cm/min). In experimental situations this species is able to climb over most obstacles placed in its path, such as a fence 2.5 cm high or 7.5-cm-square chicken wire. Most urchins could withstand starvation for three weeks and remain in good condition. After four weeks they began to lose spines and get darker in colour. Some even became cannibalistic and attacked weaker urchins. A parasitic umagillid flatworm, *Syndesmis neglecta*, occurs in the gut.

References

Berger and Profant 1961; Biermann 1998; Biermann, Kessing and Palumbi 2003; Boolootian, Giese, Tucker and Farmanfarmaian 1959; Clark 1948; Giese 1961; Jackson 1912; Jensen 1974; McCauley and Carey 1967; Moore 1959a, 1959b; Mortensen 1942; Salazar 1970; Strathmann 1971, 1975, 1978; Sumich 1973; Sumich and McCauley 1973; Westervelt and Kozloff 1992. **Range:** Jensen 1974; Sumich and McCauley 1973.

Strongylocentrotus franciscanus Giant Red Sea Urchin

Description

Strongylocentrotus franciscanus is the largest of the regular sea urchins in this region (up to 19 cm test diameter), with tapering spines about two-thirds as long as the diameter of the test. Based on tagging with dyes, some specimens may live more than 100 years.

Colour: When young, they are a light fawn brown, but with age the colour of the spines darkens to a deep reddish-purple, pale violet or pale rose. Hybrids between this species and *S. purpuratus* have been reported with the purple or violet colour of *S. purpuratus* but the long spines of *S. franciscanus*. **See colour photo C-4.**

Pore pairs: Usually 8 or 9 pairs (figure 19), but juveniles may have 6 to 10.

Spines: The mamelon on which the primary spines sit has straight sides, like a plug (figure 20). The primary spines have about 100 wedges, but the number varies. A small specimen (8 mm test diameter) may have 28–30 wedges and a larger specimen (23 mm) 70–80.

Pedicellariae: Two globiferous forms – short and long stalk; valves are slender with strongly curved terminal teeth. Two tridentate forms (figure 13c) – small and large; the valves range from parallel or slightly tapering to a blade with zig-zag teeth on the edges. Ophicephalous are constricted proximally.

Taxonomic Notes: Originally *Toxocidaris franciscana*, but Agassiz (1872) transferred it to *Strongylocentrotus*. The name *franciscanus* is after type locality near San Francisco, California.

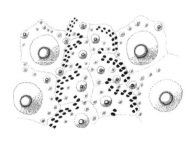

19. The arrangement of pore pairs in *S. franciscanus*.

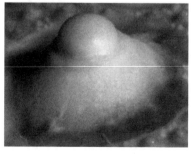

20. The mamelon of *S. franciscanus*.

Similar Species
Unlikely to be confused with other sea urchins in this region by virtue of its large size, the relatively long, tapering spines, and the reddish-purple colour.

Distribution
From Kodiak Island, Alaska, to Cedros Island, Baja California; 1–125 metres. Some researchers state that it also occurs from the Aleutian Islands to Japan, but Bazhin (1998) disputes this; he examined thousands of echinoid specimens from the western Pacific and none were *S. franciscanus*. RBCM collection 0–18 metres.

Biology
Feeding: Giant Red Sea Urchins feed mainly on drift macroalgae within kelp beds, but when these are not available they eat red foliose and attached brown seaweeds. They prefer rocky substrates but are capable of moving across sand to locate rocky areas where kelp grows.

Reproduction: Sexes are separate; females have yellow gonads and males are yellow-orange when in prime condition. During starvation both gonads become dark brown. There is no evidence of alternating sexes. In southern British Columbia, the gonads of this urchin ripen between March and September, and spawning usually peaks in May and ends by late June. In September the gonad begins to recover rapidly and reaches 80 per cent of maximum levels by October. During this time, the urchins feed primarily on drift algae released by storms, which they store as glycogen in the gonad. Between October and April the gametes develop in preparation for the next spawning peak. This is the period when people harvest the urchins for their roe. When spawned, eggs are 130–140 µm in diameter. Metamorphosis from echinopluteus larva (figure 16) to juvenile occurs 40 to 152 days after fertilization, depending on the water temperature. At settlement the juvenile is about 350 µm in diameter.

Other Notes: Juvenile Giant Red Sea Urchins are usually found under the canopy of adult spines or adjacent to adults; only one-third of the juveniles were found away from adults. Experiments show that this distribution is the result of migration to the adults rather than selective predation of unprotected juveniles or selective settlement by larvae near adults. Juvenile urchins also tend to congregate, so that some adults have no juveniles beneath them while others have many. This behaviour helps to protect juveniles from predators and may also allow them to share seaweed snared by the adults. In California, where predators of juveniles, such as spiny lobsters and certain fish, are more plentiful than in this region, virtually all the juveniles found are congregated under the spine canopy.

A study done in California showed that *Strongylocentrotus franciscanus* responds to the predatory Sunflower Star (*Pycnopodia helianthoides*) by standing its ground and bending its spines across the sea star's arm in a pinching movement. If the sea star persists, the urchin will eventually move away rapidly. Only occasionally will it use its pedicellariae to pinch the sea star's tube feet. The Sunflower Star is attracted equally to the Purple Sea Urchin and the Giant Red Sea Urchin, but the latter is significantly more successful in avoiding predation. Sea Otters (*Enhydra lutra*) prey heavily on these sea urchins where they are abundant, allowing kelp to flourish and provide sheltered habitat for fish, marine mammals and invertebrates. In areas without Sea Otters, Giant Red Sea Urchins can remove most of the seaweed and attached animals to create an urchin barren, leaving only the few species inedible to them (but see the discussion on urchin barrens in the Green Sea Urchin account, page 37).

The amphipod, *Dulichia rhabdoplastis*, is often associated with *S. franciscanus*. It attaches bits of detritus and fecal material to the tips of the urchin's longest spines, producing a filamentous strand up to 4 cm long. The surface of the strand contains a rich growth of diatoms that the amphipod consumes. During the winter there are no diatoms so the amphipod lives on detritus and captured plankton. If disturbed, it can swim effectively for several metres and will settle down and attach to any nearby sea urchin. The sea urchin does not appear to benefit from the association.

Fishery

The roe of Giant Red Sea Urchins (like that of Green Sea Urchins) is a delicacy in Asian cuisine. A commercial fishery began in California in 1973 with a catch of about 3600 tons; this rose to 24,000 tons in 1986, then dropped down to about 5000 tons by 2005. This boom-and-bust cycle, which seems typical of urchin-roe fisheries, may have been exacerbated by the increasing population of Sea Otters. (In his days as a starving graduate student, Austin harvested these urchins in marine waters off Monterey, California, and sold them for research purposes, but he also had to compete with Sea Otters that have moved into the area.)

Giant Red Sea Urchins have been harvested commercially in British Columbia since the 1970s. The harvest increased rapidly in the early 1980s on the south coast, but after 1992, it was reduced and stabilized by quotas. The fishery is well regulated in BC, with a harvest rate of 2 to 3 per cent of the estimated population allowed each year, as of 2004. In 1999, there were 110 licensed vessels with a combined quota of 5601.6 tons, 19.1 per cent allocated to the south coast and 80.9 per cent to the north coast. In 2003, the fishery harvested 4400 tons, for which the fishers netted $7.7 million and wholesalers sold for $16.4 million.

References

Breen, Carolsfeld and Yamanaka 1985; Bernard 1977; Durham, Wagner and Abbott 1980; Estes, Duggins and Rathbun 1989; McCloskey 1970; Laur, Ebeling and Reed 1986; Moitoza and Phillips 1979; Strathmann 1987; Mortensen 1921; Ebert 1998; Lawrence 1975. **Range:** Bazhin 1998; Jensen 1974.

Strongylocentrotus pallidus   White Sea Urchin

Description

A regular urchin, similar in general shape to *Strongylocentrotus droebachiensis* but with fewer spines. Test diameter up to 90 mm with a spine length of about one-quarter of the test diameter.

Colour: The general impression of the live animal is white compared to *S. droebachiensis*, but on closer examination the spines are pale green at the tips and reddish or brown at the base. The rows of tube feet are white. The test varies from pale green and greenish-brown to reddish-purple. **See colour photo C-7.**

Pore pairs: Six, occasionally seven or eight, seldom five.

Spines: The primary spines of specimens with test diameter of less than 10 mm have 20 to 24 wedges in cross section, but older ones have only 8 to 12. The mamelon of the primary tubercles has an undercut neck.

Pedicellariae: Globiferous similar to those of *S. droebachiensis* with a short neck and a globular head (when the animal is alive).

Tridentates vary from slender to wide. The slender ones have a long, narrow blade (about three times longer than the basal part); the wide type is broad-leaved and the blade is about twice as long as the base. The head is conical in shape. Ophicephalous pedicellariae are similar throughout the genus, so not diagnostic.

Taxonomic Notes: Originally described as *Toxopneustes pallidus* Sars. Some specimens of this species collected off Oregon were referred to as *Strongylocentrotus echinoides* Agassiz and Clark, 1907. The name *pallidus* is from the Latin, meaning "pale".

Similar Species

Can be confused with *Strongylocentrotus droebachiensis*, but the colour usually separates them. *S. pallidus* has 20–24 wedges in the spines and six pore pairs versus 23–27 wedges and five pore pairs for *S. droebachiensis*. *S. pallidus* is generally found in deeper water.

Distribution

Common on the Arctic coasts. Similar distribution to *Strongylocentrotus droebachiensis*, but also occurs on the east coasts of Kamchatka and Greenland; in the Pacific on the Asiatic coast to Korea (38°N) and on the North American coast to Oregon (44°N); in the Atlantic to Massachusetts Bay, Iceland, Shetland Islands and Norway; 5–1600 metres. In the Arctic it is most common at 50–150 metres on clay, shells, gravel and stones with the seaweeds *Laminaria*, *Fucus*, *Desmarestia* and red algae, and also on substrates with bryozoans and sponges. In this region it occurs at 9–490 metres with mean depth of 122 m, based on RBCM specimen records.

Biology

Feeding: Most studies indicate an omnivorous diet. On the Grand Banks off Newfoundland gut contents were single-celled benthic organisms, such as foraminiferans and diatoms, and the remains of various animals such as barnacles, bryozoans, hydroids and amphipods.

Reproduction: At Friday Harbor, San Juan Island, larvae appear in the plankton about mid March; eggs are 155–170 μm in diameter. Early larvae are very similar to *Strongylocentrotus droebachiensis* but according to Richard Strathmann, they are distinguishable by their orange pigmentation. The larvae pass from fertilization through metamorphosis and settling in 63 days. *S. pallidus* and *S. droebachiensis* can hybridize.

Other Notes: The larvae of *S. pallidus* are less tolerant of low salinity than *S. droebachiensis*. This could relate to the adult distribution of Green Sea Urchins being shallower and in more coastal areas than the deeper-water White Urchin. Studies in the Barents Sea recorded a mean density of 3.6 individuals per square metre and a maximum of 25.5. Using growth lines on the plates, an individual with a test diameter of 45 mm was estimated to be 45 years old.

References
Bluhm, Piepenburg and Von Juterzenka 1998; Falk-Petersen 1983; Gilkinson, Gagnon and Schneider 1988; Roller and Stickle 1985; Strathmann 1978, 1979, 1987; Swan 1962. **Range:** McCauley and Carey 1967; Jensen 1974.

Strongylocentrotus purpuratus Purple Sea Urchin

Description
Strongylocentrotus purpuratus is a small to medium-sized (50–100 mm test diameter) regular sea urchin with even-length, stout spines, but variable in length from one quarter to one third of the test diameter.

Colour: Juveniles start out green but change to bluish-purple as adults. Hybrids between *S. purpuratus* and *S. franciscanus* can occur, usually with the colour of one species and the spine characteristics of the other. **See colour photo C-5.**

Pore pairs: Normally a maximum of eight, sometimes nine (figure 21); animals with a test diameter of 5 mm have six pairs, and those 9–11 mm in diameter have seven pairs.

Spines: The mamelon of the primary spines is not undercut but more like a plug. Young urchins less than 10mm in diameter have primary spines with 50–60 wedges.

Pedicellariae: The globiferous type is small with slender valves. Tridentates have three

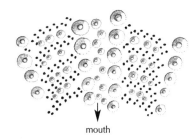

mouth

21. The pore pairs in the ambulacrum of *S. purpuratus.*

forms: small and narrow-leaved, large and broad-leaved, and large and narrow-leaved with a large basal part. Two ophicephalous forms: one similar to that of *S. pallidus* and the other with a big, round basal part and a short blade.

Taxonomic Notes: Originally *Echinus purpuratus* Stimpson, but transferred to *Strongylocentrotus* by Agassiz (1872). The name *purpuratus* is from the Latin *purpura*, meaning "purple".

Similar Species

The bluish-purple colour and the stout spines are fairly distinctive. None of the other common sea urchins have a similar colour, except perhaps the Giant Red Sea Urchin, but the length of the primary spines will separate them. The Purple Sea Urchin is usually restricted to the rocky intertidal or shallow subtidal zones of the exposed coast.

Distribution

Sitka, Alaska, to Cedros Island, Mexico; juveniles have been recorded as deep as 161 metres but adults normally live from the low intertidal zone down to about 30 metres. RBCM collection 0–24 metres.

Biology

Feeding: Omnivorous and opportunistic, the Purple Sea Urchin usually feeds on attached or drifting seaweeds, but also on encrusting organisms and on detritus brought into their burrows. It snares drifting items with its tube feet. It transfers pieces of seaweed to the mouth and holds stones and shells for a few days before discarding them.

Reproduction: Ripe specimens occur on the outer coast of Juan de Fuca Strait from December to May, but most spawn in April. Females typically shed three to six million eggs; if well fed, they can shed this volume two or three times in a two-month interval. Eggs are a translucent yellow-orange and 78–80 μm in diameter. North of Monterey, California, larvae require 63 to 86 days from fertilization to metamorphosis in temperatures less than 15°C. Urchins can begin gamete production at two years old and 24 mm in diameter; most urchins over 40 mm are reproductive.

The Purple Sea Urchin has been used extensively by biologists to elucidate embryological principles and gene regulation. The species is relatively easy to obtain, survives well in captivity and can be stimulated to spawn on cue, making it a suitable lab animal. It has been designated as a high-priority organism by the National Human Genome Research Institute because of its close relationship to vertebrates.

Other Notes: When attacked by the Sunflower Star (*Pycnopodia helianthoides*) a Purple Sea Urchin spreads open its globiferous pedicellariae and moves away as quickly as it can; but the Sunflower Star can move faster. Many Purple Sea Urchins live in depressions or burrows in rocky intertidal areas. Biologists generally agree that generations of urchins have excavated the burrows, eroding the rock by the action of their teeth and spines on its surface. Most urchins stay in their burrows and rely on drifting algae for food, which they snare with their tube feet. They also defend their burrows against other sea urchins by partially emerging and pushing the intruder away without using their pedicellariae. The isopod *Colidotea rostrata* clings to the spines of this sea urchin, and the intestine may be occupied by up to 12 protozoan species.

Fishery
The roe of the Purple Sea Urchin is reportedly very similar to that of highly desirable domestic Japanese species, but so far there have been only limited fisheries in California, Oregon and British Columbia. Without proper restraints, this species could be rapidly fished out.

References
Jensen 1974; de Ridder and Lawrence 1982; Durham, Wagner and Abbott 1980; Moitoza and Phillips 1979; Strathmann 1987; Maier and Roe 1983; Ricketts, Calvin and Hedgpeth 1985. **Range:** McCauley and Carey 1967; Clark 1948.

Family Dendrasteridae

Medium-sized to large sand dollars. Well-developed petals, the anterior one more widely open than the others, which are paired; four genital pores; food grooves divide into two or three branches; anus (periproct) near the margin.

Dendraster excentricus Pacific Sand Dollar / Sea Cookie

Description

A flat, round type of urchin up to 10 cm in diameter with tiny spines like fuzz, covering the surface. The mouth is central and the anus is near the posterior edge on the oral side. The petal-like pattern on the dorsal surface corresponds to the five ambulacra of regular urchins. In this species it is offset from the centre toward the posterior. The anterior petal is longer than the others (figure 22). The food grooves are concentrated on the posterior half of the test; they branch out from the mouth distally and have many smaller side branches. See colour photo C-13.

Colour: Blackish-purple, dark brown or grey. The posterior half is usually darker than the anterior which is usually buried in the sand. Dead animals that have lost their spines are white. **See colour photos C-12 and C-13.**

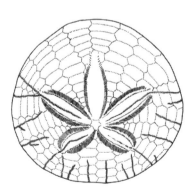

22. The dorsal side of the test of *D. excentricus*, showing petals of ambulacra.

Pore pairs: Each petal on the dorsal surface is made up of two rows of about 50 pore pairs (figure 22) on each side of the petal. The size of the petals relative to each other and their positions on the test are important taxonomic characters.

Spines: The primary spines on the ventral surface, which the Pacific Sand Dollar uses to move itself, are straight and about 4 mm long. The secondary spines are about 1 mm long. Shorter primaries occur in bands of 5 to 10, parallel to the food grooves. Tube feet occur mostly in the food grooves and on each side. The primary spines on the dorsal side are slightly bent with expanded club-shaped ends.

Pedicellariae: The tiny biphyllous type has two valves 50 μm long with fine teeth, a neck twice as long and a stem about 120 μm long. Bidentate types are similar but much larger, with valves 150 μm long and armed with three dagger-like teeth at the tip. The valves are not attached directly to the calcareous rod but to a flexible neck filled with a mucus-like substance. These pedicellariae are reputed to have poison glands but the poison has not been identified.

Taxonomic Notes: Originally *Scutella excentrica* Eschscholtz. Transferred to *Dendraster* by Agassiz and Desor, 1847. The name *excentricus* refers to the off-centre floral pattern of tube feet on the dorsal side.

Similar Species

The only other sand dollar in this region is *Echinarachnius parma,* which has the petal-like pattern in the centre of the test rather than toward the posterior edge. The food grooves of *E. parma* are symmetrical and relatively simple, with five main branches from the mouth that divide into three branches near the outer edge.

Distribution

Usually found along sheltered shores on sandy beaches near the low tide mark. Juneau, Alaska, to northern Baja California; 0–90 metres. In the Strait of Georgia and Puget Sound they tend to be intertidal and shallow subtidal (RBCM collection 0–9 metres), but on the more exposed outer coast they live deeper.

Biology

Feeding: *Dendraster excentricus* is primarily a suspension feeder, rather than a deposit feeder like most other species of sand dollars. With the body vertical in the sand, they can capture larger suspended particles and some active prey. Barrel-tipped tube feet extend outside the spines to capture dinoflagellates, small crustaceans, diatoms and pieces of algae. Tube feet beside and in the food grooves move the particles toward the mouth with the aid of mucus; cilia are not involved in moving the food. Near the mouth, oral tube feet and oral spines push the food into the mouth where it is collected and macerated by the jaws.

Reproduction: In the San Juan Islands, spawning usually occurs from mid April to July, but potentially from late March to late summer. Eggs are pale orange, 110–125 µm in diameter and covered by a 60–80 µm jelly coat. Advanced plutei larvae are present between July and October. Metamorphosis can occur in three to eight weeks, but usually five to six weeks, at 11–14°C, when the larvae are about 750 µm long.

Other Notes: *D. excentricus* can occur in densities as high as 629 per square metre in sand. The growth rate is fairly steady until about the fifth year, when it slows greatly. Animals of five to nine years are similar in size. Few live longer than nine years. Age can be determined by counting growth rings on the plates after they have been cleaned and then soaked in cedar oil or olive oil. Adults can bury themselves in about 15 minutes and are found as deep as 10 cm below the surface.

When exposed by the tide, sand dollars bury themselves, but at high tide the posterior half protrudes from the sand in a semi-vertical position. The primary spines on the oral side provide most of the locomotion, while the anterior marginal spines are responsible for the vertical digging action. The forest of primary spines and smaller miliary spines prevents most of the sediment from fouling the surface of the test, but any particles that get through are whisked away by currents created by cilia on the stems of the spines. A complex pattern of currents moves the particles away from the respiratory areas and off the edge of the test.

Few predators are known, except for the Glaucous-winged Gull (*Larus glaucescens*), which can break the tests of exposed individuals and eat the soft parts. The Pacific Sand Dollar's distribution in the intertidal zone greatly reduces predation by the Sunflower Star (*Pycnopodia helianthoides*), Vermilion Star (*Mediaster aequalis*) and Giant Pink Star (*Pisaster brevispinus*), all of which eat *Dendraster* but live in adjacent subtidal areas.

The flatworm *Syndesmis dendrastrorum* occurs in the intestine anywhere from the esophagus to the rectum.

References

Birkeland and Chia 1971; Chia 1969a, 1969b; Durham 1955; Durham, Wagner and Abbott 1980; Mooi 1986a, 1986b, 1986c, 1989, 1997; Mortensen 1948; Strathmann 1987; Timko 1976. **Range:** McCauley and Carey 1967.

Family Echinarachniidae

Medium to large sand dollars. Anterior petals more open than paired petals; four genital pores; anus on the margin; food grooves have a central trunk.

Echinarachnius parma   **Northern Sand Dollar**

Description
Flat and circular, up to 80 mm in diameter, with a velvety surface of tiny spines.

Colour: Greyish-brown, brownish-red, flesh red or lilac; spines greenish or brown. **See colour photo C-10.**

Pore pairs: The petals (about three-fifths of the radius) form a symmetrical pattern in the centre of the test (figure 23). The petals are more open at the end than *Dendraster excentricus*. On the oral surface, five straight food grooves radiate from the mouth and divide into three near the periphery of the test (about two-thirds of the radius). The anus is on the posterior edge of the test.

Spines: Five types. Club-shaped spines on the dorsal surface have an expanded end that protrudes above the smaller miliary spines that form a second canopy below that of the club-shaped spines. Fringe spines around the perimeter of the test are large and slightly flat; they constantly agitate the substrate to push sand onto the dorsal surface. Locomotory spines about 1 mm long, found all over the oral surface except in the food grooves, help to move the animal through the sand. Large circumoral spines (2.3 mm) close over the mouth when not feeding.

Taxonomic Notes: Originally *Scutella parma* Lamarck. In the Latin, *parma* means "small shield".

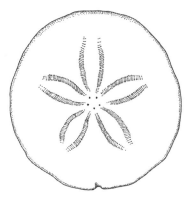

23. The test of *E. parma*. See also colour photo C-9.

Similar Species

Dendraster excentricus is the only other species of sand dollar in this region and the most abundant and common. It is usually a purplish-black and oval or elliptical in shape; but to be sure, check the position of the pore petals, which unlike those in *Echinarachnius parma*, are off-centre with the anterior petals being much longer than the posterior ones. The food grooves on the oral surface are mostly concentrated on the posterior half of the test and they give off irregular branches between the mouth and the perimeter, like a small river system. In *E. parma* the five food grooves radiate evenly from the mouth and split into three branches near the edge of the test.

Distribution

In eastern North America from Labrador to Chesapeake Bay; in northeast Asia from the Bering and Chukchi seas to Kamchatka, Sakhalin and northern Japan; in western North America from Point Barrow along the Aleutian Islands to Prince William Sound. It is primarily an intertidal or shallow-water species, but Baranova indicated that it occurs down to 150 metres and Mortensen showed that one subspecies lives down to 1625 metres, but this latter depth should be viewed with skepticism.

Although several references indicate that this species occurs as far south as Puget Sound, we have been unable to locate any original literature that refers to an actual specimen in this region. We have never seen it or collected it in the waters of British Columbia. In Sitka, Alaska Lambert found *Dendraster excentricus* to be the common species. Mortensen (1948) lists the fossil subspecies *Echinarachnius parma sakhalinensis* as occurring south to Puget Sound. Perhaps that reference has been misinterpreted as a living species. But it is possible that *E. parma* occurs in southeast Alaska in small numbers and for that reason we include it here.

Biology

Feeding: On the east coast of North America it is known to feed on a variety of particulate organic matter smaller than 100 μm in diameter, such as organic detritus, diatoms, nematodes, ostracods, algal filaments and sponge spicules. The dense covering of primary and miliary spines keeps out most of the sand grains that could foul the surface of the test. The shafts of these spines possess cilia that create a current from the apex to the edge on the dorsal side and from anterior to posterior on the oral side. Particles that penetrate the dense thatch of spines are moved by these currents to the edge of the test and discarded. Long, barrel-tipped tube feet pick up food-laden particulates and transfer them to the tube feet in the food grooves, which then move them to the mouth.

Reproduction: A study in Maine documented maximum gonad size in August and minimum in February. *Echinarachnius parma* spawned in late November and early December after being placed in holding tanks.

Other Notes: Life span is estimated at 21 years in the Sea of Japan and more than 15 years in the Middle Atlantic Bight, off New York.

References

Ellers 1985; Ellers and Telford 1984; Ghiold 1982; Harold and Telford 1982; Mooi 1986a, 1986b, 1986c; Mooi and Telford 1982; Mortensen 1948; Steimle 1990. **Range:** Baranova 1966; Durham 1955.

Family Schizasteridae

Test usually heart-shaped, depressed in front and truncated at the back; three anterior, petal-shaped ambulacra are sunk into the test, and the two posterior petals are short; the anus (periproct) is on the upper posterior end and the mouth (peristome) markedly anterior; primary spines fairly uniform and curved backwards.

Brisaster latifrons   **Heart Urchin**

Description

Brisaster latifrons is an irregular sea urchin that is roughly egg-shaped and bilaterally symmetrical, with the mouth at one end rather than on the underside. It grows up to 73 mm long. Among the spines are five ambulacra (petals) on the dorsal surface; the three anterior ambulacra are two to three times longer than the two posterior ambulacra (figure 24). The spines covering the test are flat and curved at the ends. A thin, black line, the *fasciole*, consisting of small spines with knobs on the end, traces a wavy pattern around the ends of the ambulacra. **See colour photo C-11.**

Taxonomic Notes: Originally *Schisaster latifrons* Agassiz. There is much debate about the difference between this species and *Brisaster townsendi*. McCauley (1967) synonymized the two species, but more recently, Hood and Mooi (1998) showed that they could be separated on the basis of the posterior petal width. The ranges of the two species overlap in southern California, so it is possible that *B. townsendi* and *B. latifrons* are hybridizing and causing confusion. The name *latifrons* is from the Latin *latus*, meaning "broad", and *frons*, meaning "leaf", referring to the petal-like pattern.

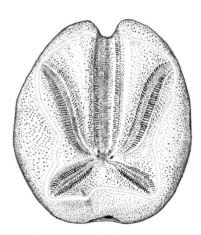

24. The test of *B. latifrons*.

Similar Species

In southern California, *Brisaster townsendi* is similar in general appearance. The main difference between them is the width of the posterior petals: those of *B. townsendi* are long and wide, while those of *B. latifrons* are short and narrow; but *B. townsendi* has not been found in the region covered by this book.

Distribution

From the Aleutian Islands, Alaska, to southern California and the Gulf of California. Usually found in deep water where the sediments are deep and soft; RBCM collection 51–1166 metres. The type specimen was collected off Baja California at 1800 metres depth.

Biology

Feeding: Heart Urchins are deposit feeders, ingesting mud as they move through it. They live below the mud surface but maintain a chimney to the surface with large modified tube feet to obtain oxygen. Once, while using a remotely operated vehicle, Lambert noticed a line of holes in the soft mud. Excavation revealed a *Brisaster latifrons* a few centimetres below the surface – it must have been moving slowly through the mud and creating successive respiratory chimneys.

Reproduction: In the Strait of Georgia and Puget Sound, this species spawns yellow-green eggs (330–355 µm) in March. The echinopluteus larvae (similar in appearance to those of *Strongylocentrotus franciscanus* – see figure 16) have orange-red pigment spots and arms about a millimetre long. They feed on plankton, then settle to the substrate 67 to 167 days after fertilization. It is unusual that such a large egg develops into a feeding larva; if necessary, the larva of this species can grow and settle without feeding.

Other Notes: Few predators have been documented. A *Brisaster townsendi* was reported to be eaten by a deep-sea star *Rathbunaster californicus* and another was found in the stomach of a Giant Pacific Squid (*Moroteuthis robustus*). Off Oregon, McCauley recorded mean densities from 0.2 to 3.8 individuals per square metre with the use of an anchor dredge. In the upper Santa Barbara Basin, off southern California, Thompson recorded *Brisaster latifrons* in mean densities as high as 30 per square metre.

References
Hart 1996; Hood and Mooi 1998; McCauley 1967; Nichols 1975; Strathmann 1978; Strathmann 1979; Talmadge 1976; Thompson, Jones, Laughlin and Tsukada 1987. **Range:** McCauley and Carey 1967; Mortensen 1951.

COLOUR PHOTOGRAPHS

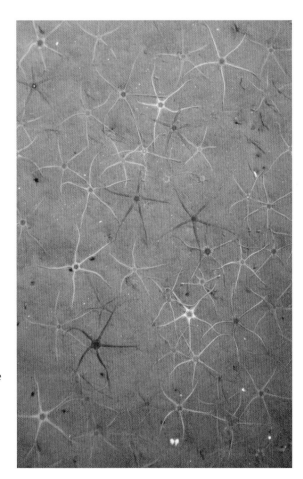

C-1. A dense bed of *Ophiura sarsii* on the muddy ocean bottom. (C-20 shows a closer view of one animal.)

C-2. Common Feather Star, *Florometra serratissima*. Page 16.

C-3. The mouth region of *F. serratissima*.

C-4. Giant Red Sea Urchin, *Strongylocentrotus franciscanus*. Page 40.

C-5. Purple Sea Urchin, *S. purpuratus*. Page 45.

C-6. Green
Sea Urchin,
*Strongylocentrotus
droebachiensis*.
Page 34.

C-7. White Sea
Urchin, *S. pallidus*.
Page 43.

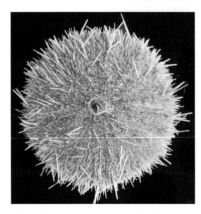

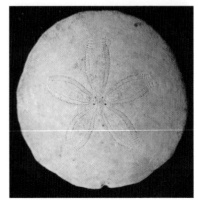

C-8. Pink Sea Urchin, *S. fragilis*.
Page 37.

C-9. Dorsal test of a Northern Sand
Dollar, *Echinarachnius parma*.

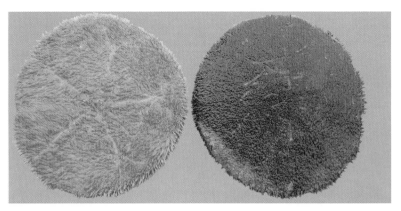

C-10 (above). Ventral (left) and dorsal views of *E. parma*. Page 51.

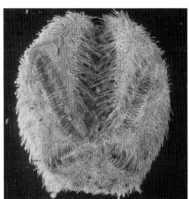

C-11 (right). Heart Urchin, *Brisaster latifrons*. Page 54.

C-12. Pacific Sand Dollar, *Dendraster excentricus*. Page 48.

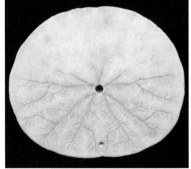

C-13. Ventral test of *D. excentricus*, showing the mouth and food grooves

C-14 (above). Basket Star, *Gorgonocephalus eucnemis*. Page 73.

C-15 (below). *G. eucnemis* feeding at Tasu Sound, Queen Charlotte Islands.

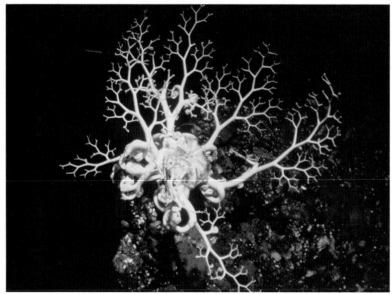

C-16. *Asteronyx loveni.* Page 71.

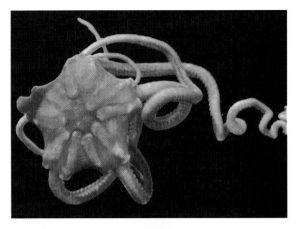

C-17. *Ophiacantha cataleimmoida.* Page 76.

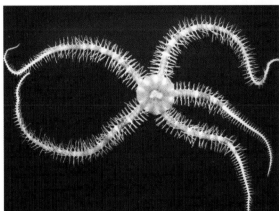

C-18. *Ophiacantha normani.* Page 80.

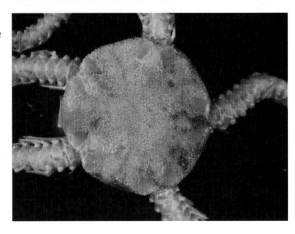

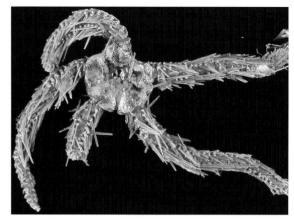

C-19. *Ophiacantha diplasia.* Page 78.

C-20. *Ophiura sarsii.* Page 86. (C-1 shows an aggregation on the ocean floor.)

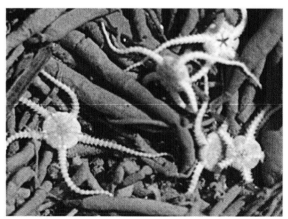

C-21. Five *Ophiura leptoctenia.* Page 82.

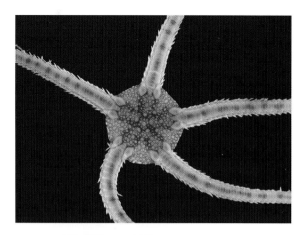

C-22. *Ophiura luetkenii.* Page 84.

C-23. Oral view of *O. luetkenii.*

C-24. *Amphiodia periercta.* Page 94.

C-25. *Stegophiura ponderosa.* Page 88.

C-26. *Amphioplus macraspis.* Page 99.

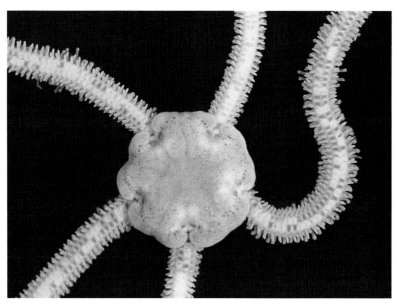

C-27. *Amphiodia occidentalis.* Page 91.

C-28. *Amphiodia urtica.* Page 96.

C-29. *Amphipholis squamata* (preserved specimen). Page 105.

C-30. *Ophiopteris papillosa* ventral view.

C-31. *O. papillosa* dorsal view. Page 120.

C-32. *Amphioplus strongyloplax.* Page 100.

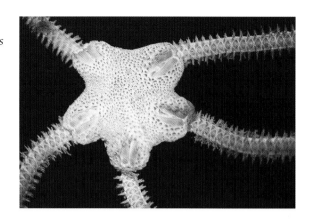

C-33. *Ophiopholis japonica.* Page 115.

C-34. *Ophiopholis bakeri.* Page 113.

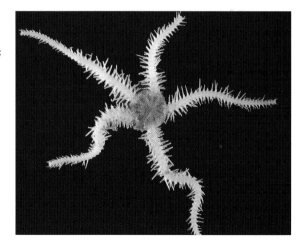

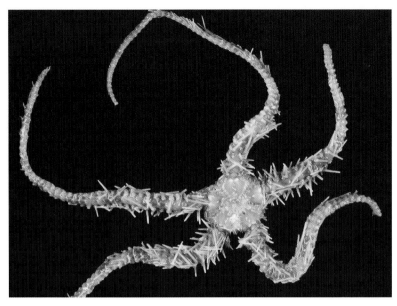

C-35. *Ophiopholis longispina.* Page 118.

C-36. *Ophioplocus esmarki* (preserved specimen). Page 123.

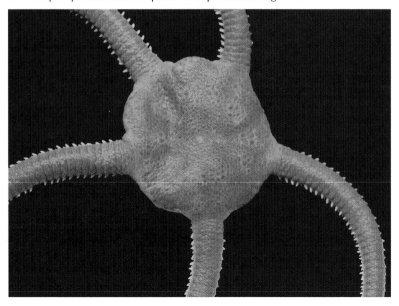

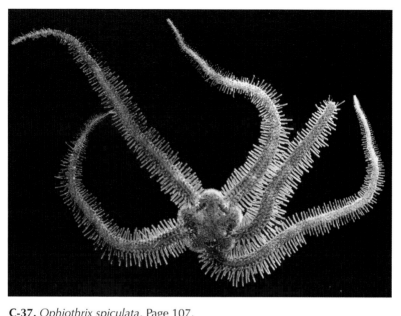

C-37. *Ophiothrix spiculata.* Page 107.

C-38. Ventral view of *O. spiculata*, showing the mouth area.

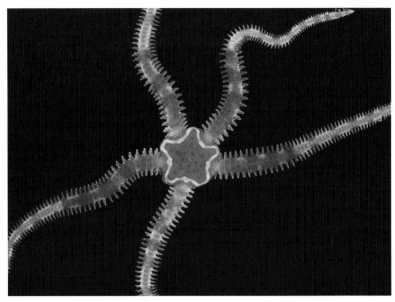

C-39 (above). Daisy Brittle Star, *Ophiopholis kennerlyi*. Page 116.

C-40 (below). Another colour pattern of *O. kennerlyi*.

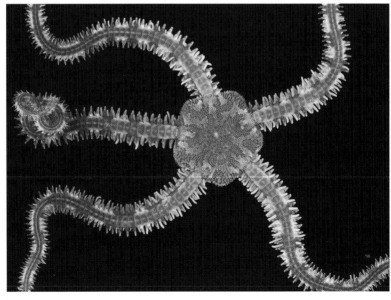

BRITTLE STARS (OPHIUROIDS)

Introduction

The first known fossil brittle stars date back to the Early Ordovician, about 500 million years ago. Today, this group, class Ophiuroidea, occurs in marine waters from the intertidal zone to the abyss and from the poles to the tropics, with 250 described genera and 2000 species. From southeast Alaska to Puget Sound we know of 53 species in 24 genera – of these, we include 24 species in 12 genera that live in depths of 0 to 200 metres. Some species have northern origins, extending south into our region from the Bering Sea, while others enter our region from a southern distribution. These numbers will change as we explore deeper water and discover new records.

External Anatomy

Brittle stars are named for their ability to detach an arm to escape from predators. The scientific name "ophiuroid" comes from the Greek *ophis* meaning "snake" and *ura* for "tail", hence the other common name, "Serpent Star" referring to the appearance of the typically long, sinuous arms. A typical brittle star has a central disc sharply marked off from the arms. Most species have five arms, but some have six and a few even have seven, eight or nine. In one group, the basket stars, the arms are branched.

These animals are made up of many calcareous plates either fused together or connected by ligaments that allow flexibility. The dorsal (aboral) surface of the disc, may be smooth or covered with granules or spines, above a series of scales or plates (figure 25). The

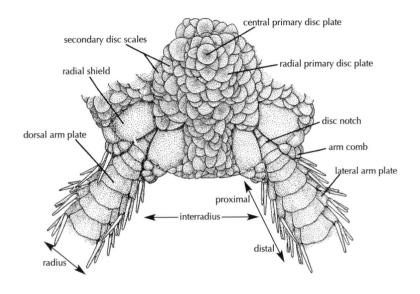

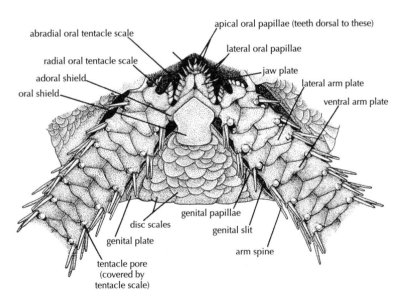

25. The external anatomy of a brittle star, dorsal above and ventral below.

arrangement of the plates is a diagnostic character. Most species have an obvious pair of radial shields where an arm meets the disc, although in some, these may be partly concealed. The arms are long, relative to the disc, and made up of articulating joints. Generally, each joint is covered by lateral, dorsal and ventral plates of various shapes, and spines of various lengths, all characteristic for a particular species. Two to fifteen arm spines form a vertical row on the lateral plate.

On the ventral (oral) surface, the arms continue into the centre and blend in to the mouth area, a star-shaped opening formed by five wedge-shaped jaws. Each jaw has one or two oral papillae at the tip. The family Amphiuridae can be distinguished by its pairs of apical oral papillae. Below the papillae the teeth form a vertical row into the mouth when viewed from the oral side. In two families, a cluster of bumps called dental papillae cover the tip of each jaw. In some families, a series of oral papillae line the sides of each triangular jaw. The arrangement and number of these papillae is an important feature – and often a point of confusion – in separating some families. Other than the jaw plates, the oral surface is dominated by the oral shield bordered on either side by the adoral shield. The shapes of these plates help identify a species. To add to the confusion, some species (e.g., *Amphioplus strongyloplax* – see figure 54) have a tentacle scale in the mouth cavity that looks very much like an oral papilla but slightly more dorsal in position, so this would not be counted in the number of oral papillae. The number given refers to the number on one side and includes the apical papilla. The shapes of the oral and adoral shields are also important characters.

In most species, on either side of the arm a deep genital (bursal) slit runs from the edge of the disc to the oral shield, and is often edged by genital papillae. These slits lead into a genital bursa, which serves a respiratory function as well as receiving gametes from the gonads. On the oral surface of each arm joint at the boundary between the oral and lateral plates, tube feet protrude from a pair of podial pores. Along the edge of each pore are one or more taxonomically important tentacle scales. In many groups the first two podial pores are inside the mouth, but in the genus *Ophiura*, for example, the second one is exposed on the edge of the jaw cavity at the base of each arm (see figure 41).

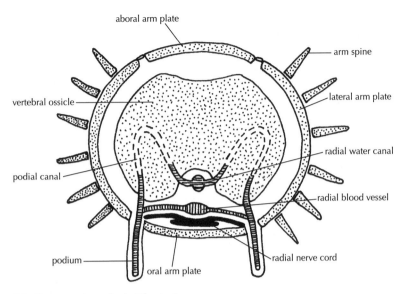

26. Cross-section of a brittle star's arm.

Internal Anatomy

Deep in the oral cavity the circular mouth leads into the esophagus and a large stomach, which fills much of the disc. Except for one family, there are no branches into the arms. Ophiuroids have no intestine or anus. A water ring around the esophagus sends a radial water canal into each of the arms, which branches laterally to supply each set of tube feet. The tube feet do not have bulb-like ampullae like those of sea stars, but each has a valve near the base. A nerve ring and a haemal ring around the mouth give off a radial nerve and a radial haemal (blood) vessel along the oral side of each arm (figure 26). Four muscles between adjacent vertebral ossicles allow the arms to flex and, in some cases, roll up.

Reproduction

Brittle stars from cold, temperate waters typically attain sexual maturity in two to three years, then breed annually. The small sac-like gonads within the body cavity attach to the wall of the genital bursa near the genital slits. There may be one or two per bursa or, as in the case of a basket star, up to several thousand. The gonad discharges ripe gametes into the bursa and they pass out through the slit. The animal often raises its disc off the substrate to allow the gametes to be dispersed by the currents. Most brittle stars have separate sexes, which cannot be distinguished externally unless the colour of the gonad shows through the body wall. There are some exceptions, where the male is a dwarf attached to the larger female.

Some ophiuroids, such as *Amphipholis squamata*, possess both male and female gonads in each bursa, and some hermaphroditic species are protandric (first male and later female). Hermaphroditic species brood their young in the genital bursae, with the largest number of brooders known from Antarctic or sub-Antarctic waters. Brooders invest more time in producing their offspring; for example, in California, *Ophioplocus esmarki* has bursae swollen with eggs in January and is still brooding young in July.

Most species studied in our region have relatively small eggs, which they shed directly into the water. Once fertilized, the eggs develop into characteristic ophiopluteus larvae within about three weeks. Skeletal rods support four pairs of arms and the body. These rods support continuous ciliated bands used for feeding and swimming. The feeding (planktotrophic) larvae remain in the plankton for 3 to 13 weeks. Only about five local species with planktotrophic larvae have been studied in detail.

Species with larger yolk-filled eggs develop into non-feeding (lecithotrophic) larvae that remain in the plankton for a week or less. They grow only two arms or develop into simple barrel-shaped larvae, called vitellariae, with distinct rings of cilia encircling the body.

Behaviour

Brittle stars often hide under rocks. If the sheltering rock moves, the brittle star quickly drags itself to another shelter. Movement may be stimulated by food or by a physical disturbance. One species has been clocked at 180 cm per minute. The tube feet do not have suckers, but in some species they secrete mucus that allows the animal to climb the vertical glass wall of an aquarium. Many species bury themselves in sand or mud or under rocks, leaving only a portion of the arms poking out; others live on the surface of mud or rocks. Brittle stars will quickly right themselves if placed on their dorsal surface. Basket stars can move to new locations to get a better feeding position. Tropical species often hide during the day and climb up to their feeding perch at night.

Many individuals show evidence of arm regeneration. As the name brittle star implies, they lose arms readily when handled, and may drop them intentionally by a process called autotomy. Nerve impulses stimulate the ligament connecting two arm segments to disintegrate. Most brittle stars can survive with only one arm remaining and grow the missing arms at a rate of about 2.3 mm per month. Others can lose their disc and gut, leaving only the mouth frame and arms, and regenerate a new disc.

Some ophiuroids produce bioluminescence. The yellowish or yellowish-green light usually emanates from the arm ossicles and not from the central disc or tube feet. It spreads from the point of stimulation by way of the radial and ring nerves. Mechanical stimulation of the disc causes the arms to glow. More studies of ophiuroid bioluminescence have been done on the wide-ranging species *Amphipholis squamata* than any other. In the Californian species *Ophiopsila californica* the flash of bioluminescence was shown to be a deterrent to visual predators such as fish and crustaceans that were startled by the sudden flash of light.

Feeding

Brittle stars are food generalists. Many can change their feeding mechanisms to either trap microscopic organisms or capture animals the size of their expanded jaws. Some feed on either plankton in the water or animal prey on the bottom. At least some can absorb soluble organic nutrients through their skin. The species accounts contain more details of feeding behaviour.

Parasites and Commensals

Parasites that attack brittle stars include ciliates, infusorians, mesozoans, polychaete worms, polyclads, trematodes, small crustaceans, snails and parasitic "barnacles". Brittle stars, themselves, can be parasites: in several species, the small parasitic male clings to the larger female near her genital slits. They can also be commensal with other marine organisms, such as sea urchins. A number of species normally cling to glass sponges and *Asteronyx loveni* is often found clinging to a sea whip.

Predators

Brittle stars are eaten by some bottom fish, such as halibut, skates and sculpins, and by several kinds of crabs and the Sand Star (*Luidia foliolata*) and other sea stars (*Pseudarchaster alascensis* and *Lophaster furcilliger*). Some species of brittle stars avoid predators by burying themselves in the sand during the day and emerging at night to forage. We know little about how species in this region avoid predators, but *Ophiura luetkenii* has a dramatic escape response when threatened by the Sand Star.

General References

A.M. Clark provides a good introduction to echinoderms in her book *Starfishes and Related Echinoderms* (1977). Nichols 1969 is a slightly more technical book about echinoderms and ophiuroids in general. For the west coast of North America, Ricketts et al. 1985 and Kozloff 1983 are useful books on intertidal life. Sept 1999, Harbo 1999, and Lamb and Hanby 2005 offer colour photographs with brief notes. Austin and Hadfield 1980 describes the California fauna, which includes species also found in British Columbia.

Some more technical literature: Hyman 1955, a standard reference for all echinoderms; Lawrence 1987 reviews the functional anatomy of brittle stars; Hendler 1991 reviews general reproduction; and Strathmann and Rumrill 1987 covers reproduction of species from the northern Pacific coast of North America. For details of internal microscopic anatomy of ophiuroids, consult Byrne 1994, and Wilkie 1988 reviews that peculiar tissue called mutable collagen.

The literature for taxonomy and classification of brittle stars is a bit more obscure and probably more difficult to find unless you have access to a university library. Mortensen 1977 (a reprint of the original book from 1927) has some useful identification keys, even though it focuses primarily on the North Atlantic. D'yakonov 1967 covers species found in the North Pacific and the seas off northeastern Asia. Clark 1911 is the most important taxonomic reference for brittle stars of the North Pacific, as it contains many of the original descriptions for the group in this area. Kyte 1969 provides one of the first and most complete identification keys for the brittle stars known in this area, and Kozloff 1987 has keys for those in shallow waters. More recently, Hendler 1996 updates the taxonomy for a number of species in this region.

Checklist of Brittle Stars

This checklist contains the names of brittle star species known from southeast Alaska through British Columbia to Washington. The numbers on the right indicate the depth or depth range in metres – the range applies to the species' entire geographic range. The classification is based on Smith, Paterson and Lafay (1995), who used morphological and molecular data to modify the older scheme of Spencer and Wright (1966). Species in bold are described in this book.

CLASS OPHIUROIDEA
Subclass Ophiuridea
Order Euryalida
    Family Asteronychidae
      Asteronyx loveni Müller & Troschel, 1842   100–4721
    Family Gorgonocephalidae
      Gorgonocephalus eucnemis Müller& Troschel, 1842   8–1850
    Family Asteroschematidae
      Astroschema sublaeve Lütken & Mortensen, 1899   605–1860

Order Ophiurida
  Suborder Ophiomyxina
    Family Ophiomyxidae
      Ophioscolex corynetes (H.L.Clark, 1911)   538–1234
  Suborder Ophiurina
    Family Ophiacanthidae
      Ophiacantha abyssa Kyte, 1982   3354–4260

Ophiacantha bathybia H.L.Clark, 1911	421–3611
Ophiacantha cataleimmoida H.L. Clark, 1911	130–1940
Ophiacantha diplasia H.L. Clark, 1911	71–1178
Ophiacantha eurypoma H.L. Clark, 1911	1041–2871
Ophiacantha normani Lyman, 1879	37–3000
Ophiacantha rhachophora H.L. Clark, 1911	366–1204
Ophiacantha trachybactra H.L. Clark, 1911	805–1143
Ophiacanthella acontophora (H.L. Clark, 1911)	419–2226
Ophiolimna bairdi (Lyman, 1883)	578–2549
Ophiosemnotes paucispina (H.L.Clark, 1911)	421– 881

Infraorder Chilophiurina
    Family Ophiuridae
    Subfamily Ophiurinae

Amphiophiura bullata pacifica Litvinova, 1971	2507–6380
Amphiophiura superba Lütken & Mortensen, 1899	51–1820
Ophiocten hastatum Lyman, 1878	824–4700
Ophiocten pacificum Lütken & Mortensen, 1899	916–1640
Ophiosphalma jolliense (McClendon, 1909)	17–1910
Ophiura bathybia H.L. Clark, 1911	2869–4425
Ophiura cryptolepis H.L. Clark, 1911	421–1280
Ophiura flagellata Lyman, 1878	128–2014
Ophiura irrorata (Lyman, 1878)	1141–3292
Ophiura leptoctenia H.L. Clark, 1911	25–3239
Ophiura luetkenii (Lyman, 1860)	0–1265
Ophiura sarsii Lütken, 1855	0–1460
Stegophiura carinata D'yakonov, 1954	846–2300
Stegophiura ponderosa (Lyman, 1878)	73–1436

    Subfamily Ophioleucinae

Ophioleuce oxycraspedon Baranova, 1954	2440

Infraorder Gnathophiurina
    Superfamily Gnathophiuridae
    Family Amphiuridae

Amphiodia occidentalis (Lyman, 1860)	0–367
Amphiodia periercta H.L. Clark, 1911	0–92
Amphiodia urtica (Lyman, 1860)	0–370
Amphioplus euryaspis (H.L. Clark, 1911)	124–582
Amphioplus macraspis (H.L. Clark, 1911)	1–876
Amphioplus strongyloplax (H.L. Clark, 1911)	40–623
Amphipholis pugetana (Lyman, 1860)	9–1204

Amphipholis squamata (Delle Chiaje, 1829)   0–1330
Amphiura arcystata H.L. Clark, 1911   15–844
Amphiura carchara H.L.Clark, 1911   110–3611
Amphiura diomedeae Lütken and Mortensen, 1899   935–2877
Subfamily Amphilepidinae
Amphilepis patens Lyman, 1879   385–4087
Family Ophiotrichidae
Ophiothrix spiculata Le Conte, 1851   0–2059

Family Ophiactidae
Ophiopholis aculeata L. 1767   0–366
Ophiopholis bakeri McClendon, 1909   18–1204
Ophiopholis japonica Lyman, 1879   15–1884
Ophiopholis kennerlyi Lyman, 1860   0–435
Ophiopholis longispina H.L. Clark, 1911   507–1253

Superfamily Ophiocomidea
Family Ophionereididae
Family Ophiocomidae
Ophiopteris papillosa (Lyman, 1875)   0–140

Infraorder Ophiolepidina
Family Ophiolepididae
Ophiomusium glabrum Lütken & Mortensen, 1899   878–5203
Ophiomusium lymani Wyville Thompson, 1873   51–2906
Ophiomusium multispinum Clark, 1911   1042–2149
Ophioplocus esmarki Lyman, 1874   0–70

Key to Families of Brittle Stars (Ophiuroids) in British Columbia, Southeast Alaska and Puget Sound*

1a. Disc and arms covered by thick skin that may contain granules, but plates and scales not covered by skin; arm spines point downwards. 2

1b. Disc and arms covered by scales or plates (may be covered by skin); arm spines placed laterally. 4

2a. Hooks on dorsal side of arms; arms forked; disc and arms bearing only small spinules or granules, or else naked.
. . . . .Gorgonocephalidae (pages 73–75)

2b. No hooks on dorsal side of arms; distal lateral arm spines are hooks. 3

3a. Gonads restricted to disc. . . . . .Asteronychidae (pages 71–72)

3b. Gonads extending to at least midway along arms.
. . . . .Asteroschematidae (page 76)

4a. Thick, soft skin covers the plates of disc and arms, but underlying plates and scales visible after drying; arm spines rough at tip. .Ophiomyxidae (page 76)

4b. Disc and arms not covered by thick skin; scales and plates easily visible although they may be partly concealed by spines or granules. 5

5a. Small, spine-like dental papillae forming a cluster at the apex of each jaw deeper in the mouth. 6

---

* Key adapted from Fell 1960.

5b. No cluster of dental papillae, just regular teeth and oral
     papillae. 7

6a. Oral papillae border each jaw. Ophiocomidae (pages 120–22)
6b. No oral papillae. . . . . .Ophiotrichidae (pages 107–9)

7a. Paired oral papillae at apex of jaw.
     . . . . .Amphiuridae (pages 91–106)
7b. An unpaired oral papilla at the apex of each jaw. 8

8a. Arms inserted laterally and firmly attached to the disc or
     fused in a notch (e.g., *Ophiura*). 9
8b. Arms inserted under the disc and not firmly fused; disc
     overhanging the arms and loosely attached. 10

9a. Granules cover the disc scales of upper and lower surfaces,
     often on jaws, too. . . . . . .Ophiodermatidae (not found in BC)
9b. No granules on plates. . . . . .Ophiuridae (pages 82–90)
     . . . . .and Ophiolepididae (pages 123–24)

10a. Margin of jaws bear uniform oral papillae (in varying
     numbers) along the side of the jaw plate. 11
10b. Oral papillae not in a continuous, uniform series; often there
     is a space between lateral oral papillae and apical papillae of
     jaw plates; dissimilar papillae. . .Ophiactidae (pages 110–19)

11a. No granules or spinules on disc. 12
11b. Granules or spinules on disc; arms slender and often
     constricted between each joint, producing a scalloped
     appearance. 13

12a. Arms robust, not constricted; a ventral keel on the midline of
     each ventral arm-plate; disc large and flat.
     . . . . .Ophiochitonidae (not found in BC)
12b. Arms slender, elongated without ventral or dorsal keels.
     .Subfamily Amphilepidinae (page 67)

13a. Arm spines numerous, long, conspicuous and erect.
     . . . . .Ophiacanthidae (pages 76–81)
13b. Arm spines few, small and inconspicuous.
     . . . . .Subfamily Ophioleucinae (page 66)

Species Accounts

The species accounts are arranged taxonomically by order and family and alphabetically within each Family. If you find the descriptions of teeth and papillae to be confusing, here is a short primer:

Turn over a brittle star and look into the mouth in the centre. You will see a star-shaped cavity lined by various tooth-like projections. Deep in the mouth there are five vertical rows of teeth on the five jaws (see figure 66 and colour photos C-23 and C-38). On the ventral surface, around the exposed edge of the jaw plates, the tooth-like projections are called oral papillae rather than teeth. Animals in the families Ophiotrichidae and Ophiocomidae have a cluster of small points called dental papillae at the jaw apex (see figure 70). Those in the family Amphiuridae have a pair of blunt apical oral papillae at the apex of the jaw plates (see figure 52). Brittle stars in other families either lack or have only one oral papilla at the tip of the jaw.

The entire geographic range and depth are given for each species so in the region covered by this book the species' depth range may be less. Based on Austin and Royal BC Museum (RBCM) collections, we have extended the known ranges for some species.

Family Asteronychidae

Moderate to large disc covered with skin or distinct scales; prominent, rib-like radial shields; long, slender arms, unbranched, skin-covered and distinctly offset from the disc; three or more arm spines, with hooks on the distal spines; small genital slit, ventral and just beyond the oral shields.

Asteronyx loveni

Description
Disc: Large, relative to arms, and covered by smooth skin, with two rib-like radial ridges from the centre to the base of each arm; diameter up to 45 mm.

Arms: Up to ten times longer than the disc diameter; long and thin and capable of coiling around objects for support. Animals greater than 9-mm disc diameter have two arms thicker and longer than the other three. Eight or nine arm spines, usually hooked, the most oral spine long and knobbed. No tentacle scales next to the tube feet.

Mouth: Four pointed oral papillae on each side of jaw and an obvious bump on the oral surface of the jaw plates.

Colour: Usually reddish-orange on the dorsal side, lighter on the ventral side; cream in preservative. **See colour photo C-16.**

Taxonomic Note: The name *loveni* is after Swedish marine zoologist Sven Ludvig Lovén, 1809–95.

Similar Species
There are no similar species in less than 200 metres depth, but deeper than 600 metres, *Asteronyx loveni* might be confused with *Astroschema sublaeve*. The disc of *A. sublaeve* is small and covered with round, rough grains and there are only two or three arm spines on each joint compared to eight or nine in *Asteronyx*.

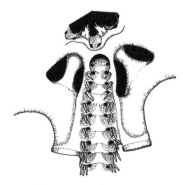

27. *A. loveni* ventral.

Distribution

A wide-ranging species with a disjunct distribution in the Atlantic and Pacific, but not in the Arctic. In the Atlantic, from the coast of Norway southwest down the coast of North America to St Vincent in the Lesser Antilles. In the Pacific, from the Bering Sea to Timor Island, Indonesia and the Indian Ocean, and in North America south to Baja California; 100–4721 metres, but usually in 300–600 metres. As shallow as 115 metres in BC (Austin collection); RBCM collection 640–1653 metres.

Biology

Habitat: When collected, this species is nearly always attached to sea whips, sea fans or other sessile animals.

Feeding: Its elevated position allows it to feed on planktonic animals in the water, and may also shelter it from bottom dwelling predators. When suspension feeding, the two thick arms tend to coil around the sea whip in opposite directions and the smaller arms extend out into the water. Mortensen speculated that *Asteronyx loveni* fed on the polyps of the sea whip; Fujita saw no evidence of this, but found fragments of crustaceans and worms in the stomach along with sediment particles.

Reproduction: Development not known, but from the large size of the egg it is likely that development is direct with no pelagic larval stage.

Other Notes: A parasitic copepod, *Chordeumium obesum*, has been found inside this species, enclosed in a sac of host tissue and displacing the gonad.

References

Clark 1911; D'yakonov 1967; Fell 1982; Fujita and Ohta 1988; Hyman 1955; Mortensen 1977. **Range:** Astrahantseff and Alton 1965; Clark 1913; Paterson 1985.

Family Gorgonocephalidae

Disc up to 14 cm in diameter with rib-like radial shields, and covered by a thick, naked skin or with scattered plates; arms vary from simple to highly branched; the arm's dorsal surface has small hooks, and the ventral surface has a furrow; oral papillae are pointed; arm spines small.

Gorgonocephalus eucnemis   **Basket Star**

Description
Disc: Up to 14 cm in diameter, indented between the radial shields and covered with thick skin. The radial shields taper toward the centre where they nearly meet with opposite shields; they are densely covered with scales bearing rough granules.

Arms: Branching and coiling at the tips, the arms are higher than their width in cross-section and have up to five lateral arm spines, four or fewer on distal branches. The spines have hooks with three teeth. Hooks are also borne on the vertical ridges of embedded plates on the dorsal and lateral arm surfaces.

Mouth: Contains spine-like teeth and oral papillae.

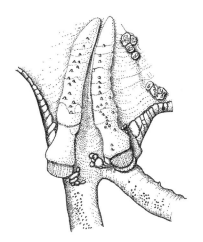 

28. *G. eucnemis* dorsal.   **29**. *G. eucnemis* ventral.

Colour: Usually tan with dark brown markings on the dorsal side of the disc, but throughout its range this species can be maroon, reddish, orange, salmon, pink and white. **See colour photos C-14 and C-15.**

Taxonomic Note: *Gorgonocephalus caryi* Verrill, 1867, originally described from the North Pacific, has been synonymized with *G. eucnemis*, described from the North Atlantic, because there are no significantly different characters to separate them. See discussion in Patent 1970 and Hendler 1996b for synonymy of *G. caryi* with *G. eucnemis*. The name *eucnemis* is from the Greek *eu*, meaning "good, well, agreeable, true", and *kneme*, meaning "shin" or "leg".

Similar Species

Gorgonocephalus eucnemis is the only species of basket star known in this region. Related species like *Asteronyx loveni*, have arms that are long and coiling but not branched.

Distribution

On the west coast of North America from Alaska to California, the Bering Sea to the Sea of Japan, Okhotsk Sea, Laptev Sea, across the Arctic to Greenland, Finmark, Spitzbergen, and south to Cape Cod; 8–1850 metres. RBCM collection 8–1240 metres.

Biology

Feeding: Juveniles of this species inhabit the pharynx of the feeding polyps of the sea strawberry *Gersemia* up to a size of 0.5 cm disc diameter. At this stage they possibly intercept the food collected by the polyp. Later they can be found attached to adult basket stars while feeding on plankton. Up to five young were found clinging to an adult. As adults they attach to the substrate with some of their arms and extend the others into the water in a dish-like shape facing the current (see colour photograph C-15). Plankton caught by hooks on the arms are rolled up in strings of mucus, and then transferred to the mouth. In the stomach, prey are wrapped in bundles of mucus. *Gorgonocephalus eucnemis* eats mostly crustaceans and arrow worms with an occasional fish embryo or jellyfish. Without cilia on the arms, this species will not likely capture unicellular organisms.

Reproduction: The sex ratio of animals around the San Juan Islands is about 1:1, with 2.6 per cent hermaphrodites. The gonads grow during the summer, and spawning occurs from October to

February. New gametes begin to form immediately after spawning, but development pauses until later in the cycle when final maturation takes place. This pause corresponds to a seasonal reduction in plankton density. Early larvae have no locomotory structures and they appear to be passively captured by *Gersemia* polyps, in which they begin their life history. Animals grow rapidly up to a disc size of 5.5 cm.

References

Austin and Hadfield 1980; Hendler 1996b; Patent 1969; Patent 1970.

Range: Clark 1911; D'yakonov 1967; Hendler 1996b.

Family Asteroschematidae

In this region, *Astroschema sublaeve* Lütken & Mortensen, 1899, occurs in deep water below 200 metres.

Family Ophiomyxidae

In this region, *Ophioscolex corynetes* (H.L. Clark, 1911) occurs below 200 metres.

Family Ophiacanthidae

A poorly defined family. Disc varies from a soft, thin skin covered with thin scales to more robust skin-covered spiny or granulated plates. Jaw plates longer than their width, with six or seven oral papillae, or equal in length and width with few papillae; arm spines usually erect, and often long and textured; tentacle pores range from large with few scales to small with several scales.

Ophiacantha cataleimmoida

Description
Disc: Up to 17 mm in diameter, rounded and flat, and densely covered by coarse, rounded granules. Radial shields rounded, naked, about one-seventh the diameter of the disc. Its general appearance is robust and spiny.

Arms: About five times as long as the disc diameter. The ventral arm plates touch; the dorsal arm plates are fan-shaped with a few granules along the outer margin, at least on the first few segments. Arm joints have six to seven long spines and a single large tentacle scale curved outward around the base of the most ventral spine.

Mouth: The jaws have one terminal oral papilla and three large oral papillae on each side. The oral shield is much wider than its length; the adoral shields do not wrap around the oral shield.

Colour: When alive, *Ophiacantha cataleimmoida* has a chocolate-brown disc and tan arms with pale spines, or a variegated tan-and-white disc and arms. **See colour photo C-17.**

Taxonomic Notes: This species was transferred to the genus *Ophiophthalmus*, but Paterson (1985) called *Ophiacantha cataleimmoida* an invalid junior synonym. We suggest that this species be retained in *Ophiacantha* for the time being. The name *cataleimmoida* is from the Greek *kataleimmos*, meaning "relic", and *eidos*, meaning "form"; Clark gave the species this nam, because according to him, it resembles the form of *O. relicta*.

Similar Species
Similar to *Ophiacantha diplasia* and *O. normani*, but *O. diplasia* has two or three long, flat tentacle scales and *O. normani* has a single sharp scale.

Distribution
Japan to the Sea of Okhotsk; Kuriles, Alaska, to northern Oregon; 130–1940 metres. RBCM collection 134–1150 metres.

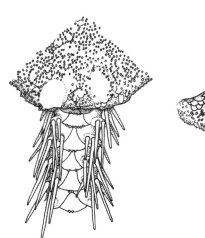

30. *O. cataleimmoida* dorsal.

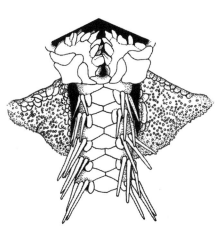

31. *O. cataleimmoida* ventral.

Biology

Habitat: On rocks, often in association with *Ophiopholis japonica.*

References

Astrahantseff and Alton 1965; Clark 1911; D'yakonov 1967; Paterson 1985. **Range:** Austin 1985; Austin and Haylock 1973; Clark 1911; D'yakonov 1967.

Ophiacantha diplasia

Description

Disc: Up to 25 mm in diameter, rounded, pentagonal and flat, and densely covered by rough, elongated granules. Radial shields are almost triangular (rounded), about one-tenth the diameter of the disc. The soft skin covering the disc often tears off during collection.

Arms: About 6.5 times longer than the disc diameter. The first ventral arm plate is u-shaped, and the proximal plates are pentagonal, then distally becoming more fan shaped with a thickened outer edge. The lateral arm plates have a prominent spine-bearing ridge. Up to seven hollow, flat, blunt arm spines, ventrally decreasing in length. The longest spine equals four to seven arm joints. Each podial pore has two long, flat tentacle scales. The dorsal arm plates are diamond- to fan-shaped with a raised distal edge.

Mouth: The jaws have one (occasionally two) terminal oral papilla with five to seven long, flat oral papillae on each side. The oral shield is oval, wider than its length and sometimes with a small lobe extending distally. Adoral shields variable but usually surrounding the lateral wings of the oral shield.

Colour: Reddish-brown disc, and brown, glassy arm spines with white ligaments at their bases. **See colour photo C-19.**

Taxonomic Notes: Kyte (1982) transferred this species to the genus *Ophiophthalmus*, but Paterson (1985) stated that *Ophiophthalmus diplasia* is an invalid junior synonym. Hendler (1996) recommended returning it to *Ophiacantha* until the genus is revised. The name *diplasia* is from the Greek *diplasios*, meaning "twice as many", signifying a pair of tentacle scales.

Similar Species

Similar in general appearance to *Ophiacantha normani* and *O. cataleimmoida,* which can be differentiated by the form of the tentacle scales. *O. diplasia* has two or three long, flat scales; *O. normani* has a single sharp scale; and *O. cataleimmoida* has a large, rounded, clearly curved scale.

Distribution

Queen Charlotte Islands to southern California; 71–1178 metres. Generally a deep water species in BC, 110–1178 metres. Clark 1911 only recorded it from 71 to 137 metres in Oregon and California. Not common in collections. Austin Collection 110–1100 metres. RBCM collection 200–1178 metres.

Biology

Habitat: Usually found on hard substrates or sessile invertebrates.
Feeding: A suspension feeder with raised arms.

References

Astrahantseff and Alton 1965; Hendler 1996b. **Range:** Austin and Haylock 1973; Clark 1911.

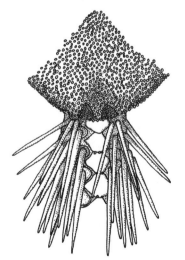 

32. *O. diplasia* dorsal.   **33**. *O. diplasia* ventral.

Ophiacantha normani

Description

Disc: Up to 22 mm in diameter, rounded and flat, and densely covered by fine, round granules. The radial shields are rounded, naked and about one-seventh the diameter of the disc.

Arms: About five times as long as the disc diameter. The ventral arm plates do not touch; the dorsal arm plates are fan-shaped with a row of granules along the outer margin, at least on the first few segments. The basal arm joints have four long spines, and the other joints have three. Each podial pore has a single pointed tentacle scale.

Mouth: The jaws have one terminal oral papilla and three long oral papillae on each side. The oral shield is much wider than its length; the adoral shields do not surround the lateral wings of the oral shield.

Colour: Disc is greyish-brown with pinkish radial shields, and the arms are tan. The oral interradial areas are brown. **See colour photo C-18.**

Taxonomic Notes: This species was transferred to the genus *Ophiophthalmus*, but Paterson (1985) called *Ophiophthalmus normani* an invalid junior synonym. We suggest that it be retained in *Ophiacantha* for the time being. The name *normani* is after Rev. Canon Alfred Merle Norman, 1831–1918, a clergyman in Durham from 1858 to 1898 who collected invertebrates by dredging. His large collection of invertebrates is now in the British Museum of Natural History.

Similar Species

Similar in general appearance to *Ophiacantha diplasia* and *O. cataleimmoida,* which can be differentiated by the form of the tentacle scales. *O. normani* has a single sharp scale; *O. diplasia* has two or three long, flat scales; and *O. cataleimmoida* has a large, rounded, clearly curved scale.

Distribution

Japan to the Sea of Okhotsk; the Bering Sea, Alaska, to Mexico. Very wide depth range: 37–3000 metres. The 37-metres record was documented by Austin and Haylock (1973) for 35 specimens at a station in British Columbia. RBCM collection 591–1910 metres.

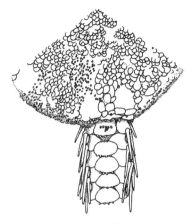 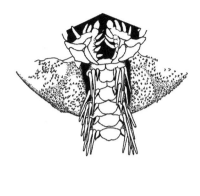

34. *O. normani* dorsal.   **35**. *O. normani* ventral.

Biology
Habitat: Rocky and muddy ocean floor.

References
Astrahantseff and Alton 1965; Clark 1911; D'yakonov 1967. **Range:** Austin 1985; Austin and Haylock 1973; Clark 1911; D'yakonov 1967.

Family Ophiuridae

Disc surrounds the base of each arm and is covered with naked scales; radial shields conspicuous; most species have a comb of papillae or spinelets on each side where the arms insert into the disc – if spinelets, they continue ventrally as papillae on each side of the genital slits; most have a single apical oral papilla, and a continuous series of pointed or rounded oral papillae on each side; the first oral tentacle pore may or may not open into the oral slit; arms usually short or moderate in length and widest at the base; arm spines often short or moderately long and folded against the arm; most species have tentacle scales.

Ophiura leptoctenia

Description
Disc: Flat and circular, up to 11 mm in diameter, and covered by 200–300 scales. The central scale and one next to the radial shields are larger than the other scales; scattered spines may be present. The length of the radial shields is about a quarter the disc diameter; they are two to three times longer than their width and usually touching at their distal ends. The arm comb has 10 or 12 long, slender, pointed spinelets, and there is often a secondary arm comb beneath it.

Arms: The dorsal arm plates are much wider than their length near the disc, becoming narrower and longer as they near the tip; they curve where they contact each other. The first ventral arm plate is about three times wider than its length, and the others become just wider than their length and diamond-shaped or elliptical. Each arm has three slender, sharp spines, the uppermost the longest and equal to or exceeding the joint. The tentacle pores are large. The first pore opens into the mouth and is protected by five or six scales on each side; the more distal pores are protected by three or four spine-like scales on the lateral plate and two or three on the ventral plate, distally reducing in number to one.

Mouth: Oral shields are longer than their width and curve distally. Six or more narrow, pointed oral papillae line each side. The genital slits are conspicuous from the oral shield to the margin of the disc. Scales along the edge of the genital slit have a series of minute papillae that grade into the arm comb.

Colour: Reddish-brown or grey and beige above; a preserved specimen is white, yellowish or pale grey. **See colour photo C-21.**

Taxonomic Note: The name *leptoctenia* is from the Greek *leptos*, meaning "fine, slender" and *ktenos*, a small comb, referring to the slender comb-like papillae.

Similar Species

The long, slender spinelets of the arm combs will distinguish this species from *Ophiura sarsii*, which has bluntly pointed spinelets, and *O. luetkenii*, which has squarish spinelets.

Distribution

A wide-ranging and common species in the North Pacific, although there seems to be a gap between Sakhalin and the western Aleutians. Ranges from Korea to the Bering Sea and from Alaska to southern California; 25–3239 metres. RBCM specimens 25–1230 metres; Austin collection 27–2000 metres.

Biology

Habitat: May be abundant on mud or partially buried in the sediment. Off central California at 1500 metres depth, researchers recorded a density of 30 individuals per 0.1 square metre.

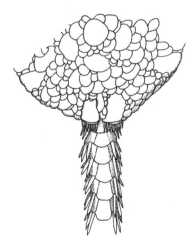

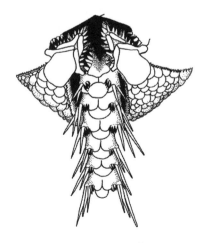

36. *O. leptoctenia* dorsal.

37. *O. leptoctenia* ventral.

Feeding: Feeds on benthic material, including small crustaceans called cumaceans.

Reproduction: Salmon-orange and pink eggs, about 0.2 mm in diameter, indicating either an ophiopluteus larva or somewhat abbreviated development.

References

Hendler 1996b; Smith and Hamilton 1983; Summers and Nybakken 2000. **Range:** Astrahantseff and Alton 1965; Austin 1985; Austin and Haylock 1973; Clark 1911, 1913; D'yakonov 1967; Kyte 1969; Nielsen 1932.

Ophiura luetkenii

Description

Disc: Up to 27 mm in diameter, pentagonal with flat, overlapping scales. The radial shields are almost pentagonal, longer than their width and about one-fifth the diameter of the disc; they are separated proximally by a wedge of scales and distally by a continuation of the dorsal arm plates. The arm combs consist of ten flat, squarish scales in tight contact with each other throughout their length.

Arms: About five to six times as long as the disc diameter. The dorsal arm plates overlap and are fan-shaped with a median ridge. The three arm spines each taper to a sharp tip; the upper spines at the base of the arm are equal to 1.5 joints in length and the others become shorter distally. The first ventral arm plate is triangular with a median groove, and the others are twice as wide as their length. Each podial pore has a small, bluntly pointed tentacle scale; large individuals have a second scale on the ventral arm plate.

Mouth: About ten spine-like apical oral papillae form a cluster around the mouth; each side of the jaw has five flat lateral oral papillae opposed by three pointed papillae on first ventral arm plate. The oral shield is a large, rounded pentagon, deeply notched on the sides. The adoral shields are narrow.

Colour: The dorsal disc surface varies but is usually mottled brown or grey. The arms often have brown or grey bands; they are white underneath. The arm spines are brown or grey at the base with white tips. **See colour photos C-22 and C-23.**

Taxonomic Note: The name *luetkenii* is after Christian Frederik Lütken, 1827–1901, a Danish zoologist and authority on ophiuroids. Originally, it was *lütkenii*, but the rules of nomenclature require that *"ü"* be replaced by an *"ue"*. The species name is sometimes published with only one *"i"*, but the original description contained two and is the correct spelling, even though modern taxonomic rules would require a single *"i"* after "Lütken".

Similar Species

Ophiura luetkenii is similar to *O. sarsii*, with which it often occurs. The form of the arm comb separates the two. *O. sarsii* has short separated spines while *O. luetkenii* has squarish spines in contact with each other throughout their length. *O. sarsii* tends to be greyish, whereas *O. luetkenii* is mottled brownish or greyish with brown or grey bands on the arms; but colour should only be used as a guideline.

Distribution

From the Bering Sea, Alaska, to Cedros Island, Baja California; 0–1265 metres, 0–366 metres in this region. Normally below 10 to 20 metres; Austin observed it intertidally north of Campbell River in a region of strong vertical mixing, where the surface temperature, salinity and nutrients were the same from the top to the bottom of the water column.

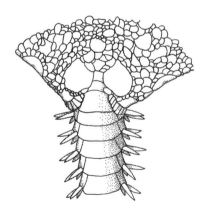

38. *O. luetkenii* dorsal.

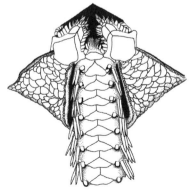

39. *O. luetkenii* ventral.

Biology

Habitat: Common on or just below the surface of silty sand or gravel substrata. Populations may be so dense that neighbours' arm tips overlap. Contact with one individual may initiate waving of the arms throughout the population.

Feeding: Limited to benthic organisms and carrion. *Ophiura luetkenii* can detect food from a distance and rapidly approach it using rowing motions of the arms. It captures food by quickly encircling it with an arm. It brings the arm loop to the jaws, which open and close as the food is pushed in by the oral tube feet. The jaws grip large pieces of flesh while the arms push it away to tear pieces off.

Other Notes: Preyed on by Dover Sole (*Microstomus pacificus*) and Copper Rockfish (*Sebastes caurinus*), as well as the Sand Star (*Luidia foliolata*). *O. luetkenii* responds to the presence of *L. foliolata* by rowing its arms to escape.

References

Austin 1966; Austin 1985; Hendler 1996b; Kyte 1969; Scouras, Beckenbach, Arndt and Smith 2004; Warner 1982. **Range:** Austin 1985; Austin and Haylock 1973; Clark 1911, 1940; Hendler 1996b; Kyte 1969; McClendon 1909.

Ophiura sarsii

Description

Disc: Up to 40 mm in diameter with rounded sides. The radial shields are roughly triangular, about twice as long as their width. The arm combs are composed of conical or triangular papillae separated at their tips.

Arms: The dorsal arm plates are short and wide; the ventral plates are triangular and usually separated by the bases of the lateral arm plates. Each arm has three spines: the upper two are 1.5 to 2 times as long as a joint; at the arm tip, the middle spine is longest. The tentacle scales are broad and flat, four or five to each podial pore, but as few as one distally.

Mouth: Four to six pointed oral papillae on each side. Narrow adoral shields partially wrap around the oral shield, which is as broad as it is long.

Colour: The disc is usually brown or grey, but may be mottled. The arms are grey or lighter than the disc. Both the disc and arms are white ventrally. **See colour photos C-1 and C-20.**

Taxonomic Note: The name *sarsii* is after Norwegian naturalist Michael Sars, 1805–69. Modern rules would add a single "i" to "sars", and the species name is often spelled that way, but the original and correct spelling is "*sarsii*".

Similar Species

Easily confused with *Ophiura luetkenii*, which co-occurs with *O. sarsii*. The arm combs of *O. luetkenii* have squarish papillae in contact with each other throughout their length; those of *O. sarsii* have pointed spines that are separated along most of their length.

Distribution

A circumpolar species: throughout the Arctic Ocean; in the Atlantic Ocean south to the Baltic Sea in Europe and Cape Hatteras in the USA; in the Pacific Ocean, from the Bering Sea to California, Japan and Korea. RBCM collection 9–550 metres; Austin collection 0–1460 metres. Very common, often in great numbers (see colour photo C-1).

Biology

Habitat: Typically on mud bottoms. Off Japan, *Ophiura sarsii* occurs in densities of $373/m^2$ and makes up 99 per cent of the macrofauna, but off Oregon there are $30/m^2$. They distribute evenly on the substrate and may or may not be in contact with neighbours.

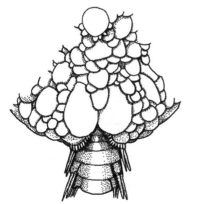
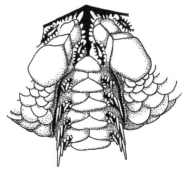

40. *O. sarsii* dorsal.   **41**. *O. sarsii* ventral.

Feeding: This opportunistic carnivore feeds on small benthic invertebrates, carrion and detritus, but it has also been observed attempting to catch larger organisms near the bottom. The distal two-thirds of the arms are raised and ready to encircle prey, such as myctophid fish, euphausids and small squid, that blunder into the ocean bottom; if a brittle star captures an animal, others join in to share the meal.

Reproduction: This species has separate sexes. In the San Juan Islands, ripe specimens collected from March to June will spawn in the lab. Orange-pink oocytes (eggs), 100–110 μm in diameter, develop into a planktonic ophiopluteus larva.

Other Notes: In dense beds of *Ophiura sarsii*, individuals avoided certain macrofauna, such as sea anemones and moving sea stars, but did not appear to avoid mobile predators, like fish and octopus. In the North Pacific, 80 per cent of the flatfishes sampled had *O. sarsii* in their stomachs. When disturbed by lights or the pressure wave of a submersible or diver, these brittle stars flee by making rowing movements with two or three arms.

References

Fujita and Ohta 1989; Jangoux 1982; Stancyk, Muir and Fujita 1998; Warner 1982. **Range:** Astrahantseff and Alton 1965; Austin and Haylock 1973; Clark 1911; D'yakonov 1967.

Stegophiura ponderosa

Description

Disc: Diameter up to 47 mm; robust, thick, heavy and stiff, with dorsal plates like paving stones. The radial shields appear swollen with a single swollen radial plate wedged between their proximal ends. The ventral side of the disc is flat.

Arms: About three times as long as the disc diameter; stiff and usually broken when collected by dredge. Triangular in cross-section, with a sharp crest along the dorsal side. Each arm has a comb where it joins the disc, consisting of two rows of small square papillae. A podial pore opens into the mouth slit and is bordered by flat tentacle scales – six or seven on the proximal side and four or five on the distal side; the next four pores are similar, but gradually reducing to two scales at arm tip. *Stegophiura ponderosa* has two types of arm spines: usually two short (less than a quarter the length of and arm joint) peg-like spines with a series of even smaller, flat, slender spines between them.

Mouth: Each side of the jaw has five or six tightly packed, squarish oral papillae and a single apical oral papilla. The adoral shields are narrow, and the oral shields are angular proximally and rounded distally.

Colour: Orange-red dorsally and white ventrally when alive; brownish when preserved in alcohol. **See colour photo C-25.**

Taxonomic Notes: Originally *Ophioglypha ponderosa* Lyman, then *Ophiura ponderosa* by Clark (1911). Revised to *Amphiophiura ponderosa* by Matsumoto (1917), then to *Stegophiura* by Kyte (1987). The name *ponderosa* is from the Latin *ponderosus*, meaning "heavy".

Similar Species
In deeper water, *Stegophiura ponderosa* could be confused with *Amphiophiura superba*, but the latter species lacks the protruding plate between the radials, and its arms are round in cross-section and have only four small arm spines.

Distribution
Aleutian Islands, Alaska, to southern California and the Sea of Okhotsk to Japan (34°N); 73–1436 metres. One record in the RBCM collection is in 73 metres in Dixon Entrance; all others are deeper

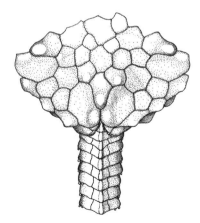

42. *S. ponderosa* dorsal.

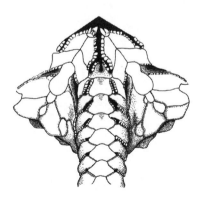

43. *S. ponderosa* ventral.

than 200 metres, with the deepest at 1436 metres, off the Queen Charlotte Islands. Usually abundant in trawls, when present.

Biology
Feeding: We know little about the feeding behaviour of this common deep-water species, but it may eat detritus, because the gut of one we examined was filled with mud.

Reproduction: Eggs 350 μm in diameter, suggesting a non-feeding larval form.

References
Kyte 1987; Litvinova and Sokolova 1971. **Range:** Austin and Haylock 1973; Clark 1911; D'yakonov 1967; Lyman 1878; Matsumoto 1917.

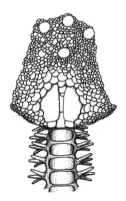 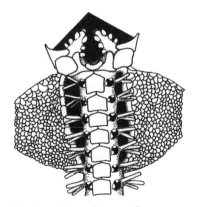

44. *Amphiodia occidentalis* dorsal.   **45**. *A. occidentalis* ventral.

Family Amphiuridae

Paired apical papillae at the tip of the jaw; disc delicate and prone to damage, usually scaled; arms long relative to the disc; arm spines short. The number and arrangement of oral papillae characterize the genera in this family.

Amphiodia

Species in the *Amphiodia* genus have approximately three equal-sized and equally spaced oral papillae on each side of the jaw. The table on the next page will help identify the local species.

Amphiodia occidentalis

Description
Disc: Up to 11 mm in diameter. The scales are fine, smooth and even. The radial shields are shaped like an elongated pear seed and separated from each other by a narrow, single row of scales. The scales on the ventral side have no spines.

Arms: Nine to fifteen times as long as the diameter of the disc. The ventral arm plates are squarish. The ventral arm spines are straight and dorso-ventrally flat. Each arm has three blunt, flat arm spines about as long as a joint. Each tentacle pore has two small, rounded tentacle scales.

Mouth: Three oral papillae on each side. The oral shields are small and shaped like a rounded diamond; adoral shields do not meet in the middle.

Colour: Grey disc with grey and white arms; specimens in alcohol are faint greenish-grey dorsally and straw coloured ventrally. **See colour photo C-27.**

Taxonomic Notes: Originally as *Amphiura occidentalis* Lyman. The name *occidentalis* is from the Latin *occidens*, meaning "of the west".

Similar Species
See the table on page 92.

Distinguishing characters for species in *Amphiodia*

Character	*A. occidentalis*	*A. periercta*	*A. urtica*
Disc	Up to 11 mm in diam.; arms 8 times disc diam.	Up to 15 mm in diam.; pentagonal.	Up to 9 mm in diam.; rounded pentagonal.
Erect scales around disc	No.	Yes.	No.
Radial shields	Twice as long as width; separated by a single row of small scales.	Twice as long as width; touching most of their length.	Three times longer than width; about 1/5 the disc diam.; most of each pair in contact.
Single or branched spines on some ventral scales	No.	No.	Yes.
Arm spines	Three stout and rounded at the end (not tapered), and dorso-ventrally flat; about as long as an arm joint.	Three, smooth and rounded, sharp-tipped, and dorso-ventrally flat; as long as an arm joint.	Three, equal in length, with a thick base tapered to a point; not flat.
Ventral arm spines under the disc	Straight.	Straight.	Curved.
Ventral arm plates	Squarish proximally, then wider than length distally; second plate pentagonal.	First plate wider than length; others squarish with a concave distal margin.	Pentagonal; length equal to width; distally on arm, becoming longer than width.
Dorsal arm plates	Oblong, width twice the length.	Two or three times wider than length.	Oval; wider than length.
Tentacle scales	Two, small and rounded.	Two, large.	Two, thin and ovoid.
Oral papillae	Three, equal sized.	Three, equal sized.	Three, unequal sized, smallest in middle.
Oral shields	Small and ovoid.	Small, pentagonal or triangular.	Pentagonal to diamond-shaped; equal length and width.

Distribution

Kodiak Island, Alaska to central California. Intertidal to 367 metres. (Hendler is suspicious of this maximum depth and suggests that records south of central California are another species.)

Biology

Habitat: In BC, populations occur two meters above to a few meters below the low-tide level in sand or shell-sand with some silt. *Amphiodia occidentalis* are often reported buried under rocks or in eelgrass beds, which may be related to the ease with which they can be pulled out of sand compared with species buried in mud. Typically, six to eight can be found under a rock, which is equivalent to about 100 per square metre.

Feeding: The arms extend up into the water and their sticky tube feet pick up suspended particles. Alternatively, the arms sweep over the surface of the ocean bottom to pick up detritus or benthic organisms. Rapid disc contractions expel waste, and then oral tube feet pass particles out.

Reproduction: A study in Monterey Bay, California, found all animals 6 mm or more across the disc to be reproductively mature. Most spawning occurred in the late spring or early summer. Eggs are yellow-green and 90–106 µm in diameter. During broadcast spawning, *Amphiodia occidentalis* elevates its disc above the substratum and disperses eggs or sperm from its bursae. Fertilized eggs develop into swimming larvae. We assume that a similar pattern occurs in our region but the timing of spawning may vary.

Other Notes: This brittle star burrows in sand by moving its tube feet up between the arm spines and piling sand above. Rhythmic contractions of the disc also aid in burrowing. The tube feet then move sand toward the arm tips, forming small mounds at the surface. Only the arm tips show above the surface. For respiration, sinuous flexing of an arm pumps water down one arm tunnel, across the disc and out another arm tunnel. This species readily sheds its arms if handled.

References

Austin and Hadfield 1980; Hendler 2007; Ricketts and Calvin 1964; Rumrill 1982; Rumrill and Pearse 1985. **Range:** Austin and Hadfield 1980; Austin and Haylock 1973.

Amphiodia periercta

Description

Disc: Up to 15 mm in diameter and covered by numerous small, irregular scales, the largest near the radial shields, which touch each other at least distally. The radial shields are about twice as long as their width. The sides of the disc are bounded by a line of erect scales. Scales on the ventral side of the disc have no spines.

Arms: Up to 45 cm long. The dorsal arm plates, two or three times wider than their length, broadly contact adjacent plates. The ventral arm plates are squarish with a concave distal margin, except for the first, which is wider than its length. Each arm has three smooth, rounded but sharp arm spines, the centre one about as long as a joint. The ventral arm spines are straight and dorso-ventrally flat. Each tentacle pore has two large tentacle scales.

Mouth: Fine scales cover the ventral interradial area. The oral shields are small pentagons or triangles, and the adoral shields are narrow, though they widen at the distal end. Each side of the jaw has three thick, rounded oral papillae.

Colour: Reddish-brown disc, which becomes fawn or yellowish-brown in preservative. **See colour photo C-24.**

Taxonomic Note: The name *periercta* is from the Greek *perierctos*, meaning "fenced around", in reference to the disc bounded by a line of scales.

Similar Species

See the table on page 92. Nielsen (1932) suggested that *Amphiodia periercta* is probably part of the variable species *A. occidentalis*, but reserved judgement until he studied more specimens. Our observations on BC populations indicate that the three species differ in morphology, size, colour, habitat and behaviour. *Amphiodia pelora* Bush, 1921, is a synonym of *A. periercta* that had also been known as *Diamphiodia periercta*.

Distribution

In North America, from the Aleutian Islands, Alaska, to central California; in Asia, from the southern Kuril Islands to the Sea of Japan; intertidal to 92 metres, but in BC, shallow subtidal down to 12 metres.

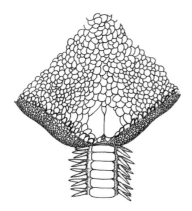 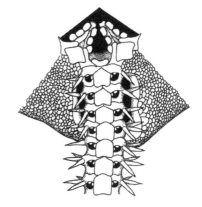

46. *A. periercta* dorsal.   **47**. *A. periercta* ventral.

Biology

Habitat: *Amphiodia periercta* usually buries itself in silty, sandy mud, but not sand. In BC, populations up to 400 per square metre inhabit shallow regions with moderate currents (25 cm/sec or 0.5 knots).

Feeding: This species eats detritus when there is no current, but as currents increase it raises one or more arms vertically and extends its sticky tube feet into the current. Small animals that adhere are passed down the arms to the mouth.

Other Notes: *Amphiodia periercta* burrows in a similar fashion to *A. occidentalis*, except that it extends its disc deeper into the substrate and its tube feet act like a conveyor belt passing substrate particles out along the burrow to the outside. It also irrigates its burrow (interpreted as respiration) like *A. occidentalis*. Cycling contractions of the disc irrigate the gonads adjacent to the sac-like bursae. A tiny pea crab dwells in the arm burrows and picks off food as it goes by. Unlike the shallow-burrowing *A. occidentalis, A. periercta* cannot be pulled out of the substrate.

References

Clark 1911; D'yakonov 1967; Kyte 1969. **Range:** Austin and Haylock 1973; D'yakonov 1966.

Amphiodia urtica

Description

Disc: Rounded to pentagonal, up to 9 mm in diameter; appears inflated; covered with small delicate scales that are larger around the radial shields. Each radial shield is one-fifth as long as the disc diameter, three times longer than its width and in contact with its neighbour for most of its length, but separated by a small wedge of scales at the proximal end. Some scales on ventral side of the disc have single or branched spines (figure 50).

Arms: Twelve times as long as the disc diameter. The dorsal arm plates are oval, wider than their length. Three equal-length arm spines have thick bases and taper to acute points; they are round but slightly compressed laterally. The ventral arm spines below the disc are curved. Two thin ovoid tentacle scales per pore. The second ventral arm plate is pentagonal, and the more distal plates are squarish with concave distal edges, about as long as their width but becoming proportionally longer farther out on the arm.

Mouth: Three oral papillae on each side: the middle the smallest and the outer the largest; all three are triangular with rounded corners, the narrow part attached to the jaw. The oral shields are pentagonal to diamond-shaped and about as long as their width; the distal ends of the adoral shields touch the first lateral arm plates. The ventral interradial area is finely scaled. The scales near the genital slits each bear a single minute clear tooth, these spiny scales sometimes extend around to the distal end of the radial shields.

 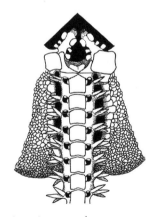

48. *A. urtica* dorsal.   **49**. *A. urtica* ventral.

Colour: Reddish to greyish-brown, with grey, pale-tipped radial shields. Specimens in alcohol are pale yellowish-brown. **See colour photo C-28.**
Taxonomic Notes: Originally as *Amphiura urtica* Lyman. The Latin *urtica* means "nettle", possibly referring to the spines on the disc's ventral scales.

Similar Species
See the table on page 92.

Distribution
Shumagin Islands, Alaska, to Mexico; intertidal to 370 metres, possibly 708 metres.

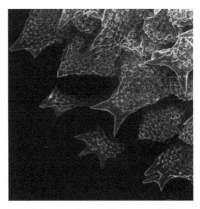

50. Spiny scales on the ventral disc of *A. urtica* taken with a scanning electron microscope (SEM).

Biology
Habitat: Buried in mud and, in BC, typically at greater depths than *Amphiodia periercta*. BC populations up to 1200 per square metre.
Feeding: Like other *Amphiodia*, this species burrows in soft sediment and feeds on detritus and suspended particles. It has been shown to be less abundant in areas influenced by wastewater and therefore useful as a bio-indicator.
Reproduction: Separate sexes, spawning in the fall in British Columbia, later in California.
Other Notes: The Lemon Sole (*Paraphrys vetulus*), Flathead Sole (*Hippoglossoides elassodon*), Sand Sole (*Psettichthys melanostictus*) and Sand Star (*Luidia foliolata*) prey on *Amphiodia urtica*. Like other burrowing Amphiuridae, *A. urtica* readily breaks off an arm tip so that a predator only gets a taste of it, then the buried disc regenerates a new arm. The top of the disc also readily breaks off; Austin observed a new disc cover regenerating itself within 14 days. In Puget Sound, *A. urtica* has a life span of about five years.

References
Austin and Hadfield 1980; Austin and Haylock 1973; Clark 1911; Hendler 1996b; Kyte 1969; Schiff and Bergen 1995. **Range:** Austin and Haylock 1973; Hendler 1996b.

Amphioplus

The *Amphioplus* genus is characterized by notched disc interradii. In collections, many specimens have lost the dorsal disc covering. Each side of the jaw has four (or rarely three) oral papillae and an oral tentacle scale between the first and second papillae; the most distal one is small. The tentacle scale gives the jaw the appearance of having five oral papillae. *Amphioplus* species in the eastern North Pacific have a vertical hole through each arm ossicle, which can be seen by removing a dorsal or ventral arm plate; this hole is absent in other amphiurid genera in the region. Local species have four to six slender, sharp arm spines and one or two tiny tentacle scales. This table will help identify the two species in this region.

Distinguishing characters for species in *Amphioplus*		
Character	*A. macraspis*	*A. strongyloplax*
Radial shields	Long (3 times width), and separated or in contact distally.	Long (4 or 5 times width), narrow and curved, and broadly in contact distally; length 1/4 the disc diam.
Oral shield	Pentagonal or hexagonal, wider than length.	Pointed proximally, rounded laterally, truncated distally, as long as width.
Ventral interradial area	Covered with scales.	Usually bare or has small, isolated scales.
Dorsal arm plates	More-or-less triangular.	Rounded and a little wider than length.
Colour	Deep red.	Light red, with some bright yellow along the arms.

Amphioplus macraspis

Description

Disc: About 10 mm in diameter, soft and swollen, and deeply notched between the arms; covered with numerous scales that are larger around the radial shields than in the centre of the disc. The radial shields are long (three times their width) and curved; separated or in contact distally. The ventral interradial area is covered with scales.

Arms: About 10 times as long as the diameter of the disc. The dorsal arm plates are almost triangular, and wider than their length. The first ventral arm plate between adjacent adorals is small, but the rest are squarish or slightly longer than their width. Each arm has five or six slender spines; the lowest is the longest and about equal to the length of one arm joint. Each arm has one or two small tentacle scales per pore near the base, and one per pore distally.

Mouth: The oral shields are pentagonal or hexagonal and much wider than their length. The adoral shields are twice as long as their width. Three or four oral papillae; the fourth, if present, is smaller and separated distally. There may appear to be five papillae, but the second is actually a tentacle scale of the first oral pore situated more dorsally in the mouth.

Colour: Red, which preserved specimens retain partly on the radial shields and some disc scales. **See colour photo C-26.**

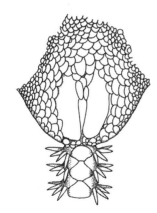

51. A. *macraspis* dorsal.

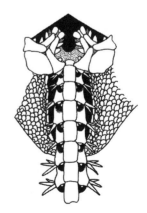

52. A. *macraspis* ventral.

Taxonomic Notes: Originally *Amphiodia macraspis* Clark, 1911; revised to *Amphioplus macraspis* by A.M. Clark (1970). Clark (1911) also described *Amphiodia euryaspis*, which, by his own account, is very close to *A. macraspis*. D'yakonov (1967) noted that Clark united *A. euryaspis* with *A. macraspis* in an unofficial communication in 1927, acknowledging the differences in the dorsal scales and the size of the radial shields, but stating that most of the other characters overlap. Pending a resolution to this question, we included *A. euryaspis* as a separate species in the checklist but did not provide a detailed description. The name *macraspis* is from the Greek *makros*, meaning "long", and *aspis*, meaning "shield", referring to the shape of the radial shields.

Similar Species
See the table on page 98.

Distribution
Widespread in the North Pacific. In Asia, from the Pacific coasts of Japan and Korea through the Sea of Okhotsk to the Bering Sea; 1–876 metres. In North America, from the Queen Charlotte Islands to Washington. RBCM specimens 1–211 metres.

Biology
Found in the stomach of Flathead Sole (*Hippoglossoides elassodon*).

References
Austin and Haylock 1973; Clark 1970; Clark 1911; D'yakonov 1967; Hendler 1996b; Ivanov 1964; Kyte 1969.

Amphioplus strongyloplax

Description
Disc: About 10 mm in diameter, soft, flat and indented between the arms. The radial shields are curved and a quarter the length of the disc diameter, four or five times longer than their width; they contact each other distally but are well separated proximally.
Arms: Long and slender, about 12 times as long as the disc diameter, and wider at mid-length. The dorsal arm plates are rounded, a

little wider than their length and scarcely in contact. Five arm spines (occasionally six) gradually taper to an acute point. The tentacle pores are large, with one or two minute scales.

Mouth: Each side has four (or five) oral papillae; the one between the first and second is actually a tentacle scale, and the distal one is tiny. The oral shields vary from pentagonal to irregularly ovoid: pointed proximally, rounded laterally, truncated distally; they are equal in length and width. The adoral shields are L-shaped and much wider at the outer ends, which insert between the first and second ventral arm plates. The oral interradial area is usually bare or has small, isolated scales.

Colour: Reddish. Austin has observed a bright yellow fluid in cavities along the arms. Preserved specimens are greyish to light brown, with orange-yellow on some arm spines and on radial and oral shields. **See colour photo C-32.**

Taxonomic Note: Originally *Amphiodia strongyloplax* Clark. The name *strongyloplax* is from the Greek *strongylos*, meaning "round", and *plax*, meaning "plate".

Similar Species
See the table on page 98.

Distribution
The Gulf of Alaska and northern British Columbia to the Mexican border; 40–623 metres. RBCM specimens 40–310 metres; Austin collection 40–373 metres.

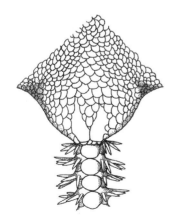

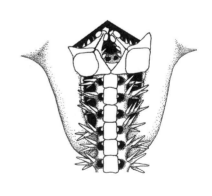

53. *A. strongyloplax* dorsal.   **54**. *A. strongyloplax* ventral.

Biology

Habitat: Muddy sand and gravel.

Reproduction: Not much known. One female had eggs 0.18 mm in diameter, which suggests a pelagic ophiopluteus larva.

Other Notes: Reported in 23 per cent of stations from Point Conception, California, to Mexico, with an overall density of 4.3 per square metre. Off the Oregon coast the estimate was 60 per square metre. *Amphioplus strongyloplax* burrows in the substrate, as evidenced by more specimens being taken in grabs (collecting devices that take a single bite out of the ocean bottom) than in trawls.

References

Astrahantseff and Alton 1965; Hendler 1996b; Kyte 1969.

Range: Austin and Haylock 1973; Hendler 1996b.

Amphipholis

The *Amphipholis* genus is characterized by three oral papillae, the most distal of which is two times wider than the others. This table will help identify the two species in this region.

Distinguishing characters for species in *Amphipholis*		
Character	*A. squamata*	*A. pugetana*
Ovoviviparous (young in bursae)	Yes.	No.
Disc-diameter to arm-length ratio	1:4–5	1:7–8 (less in small animals)
Middle arm spine-length to joint-length	Equal.	1.5 times.
Middle arm spine near arm base swollen	No.	Yes, on one side (sometimes both).
White spot on edge of radial shield	Yes.	No (rarely yes).

Amphipholis pugetana

Description

Disc: This small species has a maximum disc diameter of 5 mm; the disc scales are mostly rounded. The radial shields appear slightly sunken; they are about a fifth as long as the disc diameter and three times longer than their width; they contact each other except for a slight separation at the proximal end.

Arms: Seven or eight times as long as the disc diameter. The dorsal arm plates barely contact each other; the lateral arm plates insert between the dorsal and ventral plates; the ventral arm plates are pentagonal. Three or four moderately stout arm spines taper evenly, increasing in length ventrally; when there are only three, the middle one is longest, about 1.5 times the joint length, and swollen at the tip. Each tentacle pore has two scales.

Mouth: Three oral papillae on each side: the proximal papilla rounded and block-like, the middle one rounded but flaring from the point of attachment, and the distal one the largest, rectangular and elongated. The oral shields are shaped like diamonds or arrowheads, and about as long as their width. The adoral shields are triangular and as large as the oral shields.

Colour: Grey or orange disc, with orange and reddish-orange blotches on the arms, pale arm spines and an orange, pale brown or grey mouth area. The disc becomes greenish-grey when preserved in alcohol.

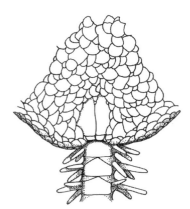

55. *A. pugetana* dorsal.

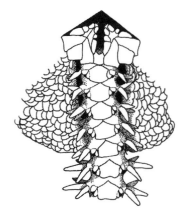

56. *A. pugetana* ventral.

Taxonomic Notes: Originally *Amphiura pugetana* Lyman. Sometimes referred to as *Axiognathus pugetana*. The name *pugetana* refers to Puget Sound, the type locality.

Similar Species

Amphipholis squamata is similar in general appearance but its arms and arm spines are shorter, and its oral shield is as long as it is wide. See the table on page 102. Because of the difficulties in distinguishing *A. pugetana* from *A. squamata*, some records are suspect.

Distribution

Gulf of Alaska to southern California; 9–604 metres. RBCM specimens 9–42 metres; Austin collection 0–1204 metres.

Biology

Reproduction: Broadcast spawning of gametes.

References

Austin and Haylock 1973; Boolootian and Leighton 1966; Clark 1911; D'yakonov 1949, 1952, 1958, 1967; Hendler 1996b; Kyte 1969; Lyman 1860, 1865; Matsumoto 1917; McClendon 1909; Ricketts and Calvin 1964; Strathmann and Rumrill 1987. **Range:** Hendler 1996b.

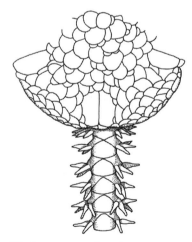

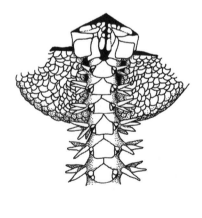

57. *A. squamata* dorsal.   **58**. *A. squamata* ventral.

Amphipholis squamata

Description

Disc: A small species, with a maximum disc diameter of 5 mm. The disc scales are mostly rounded. The radial shields are about a fifth as long as the disc diameter, and three times longer than their width; they touch along their length, except at the proximal end where they separate slightly.

Arms: Four to five times as long as the disc diameter. The dorsal arm plates barely contact each other, the lateral arm plates insert between the dorsal and ventral plates, and the ventral arm plates are pentagonal. Three or four moderately stout arm spines taper evenly; their length is equal to or less than one arm joint. Each tentacle pore has two scales: the one on the lateral arm plate is larger than the one on the ventral plate.

Mouth: Each side has three oral papillae, the proximal papilla rounded and block-like, the middle one rounded but narrow at the point of attachment, and the distal one, the largest, elongated but with a rounded outer edge. The oral shields are pentagonal with a rounded distal edge; they are about as long as their width. The adoral shields are triangular and almost as large as the oral shields.

Colour: The disc can be black, grey, beige, brown, orange or spotted; the arms are blotched with orange and reddish-orange; the arm spines are pale; and the mouth area is orange, pale brown or grey. The disc becomes greenish-grey when preserved in alcohol. **See colour photo C-29.**

Taxonomic Notes: Originally *Asterias squamata* Delle Chiaje. Sometimes referred to as *Axiognathus squamata*. The name *squamata* is from the Latin *squama*, meaning "scale".

Similar Species

Amphipholis pugetana is similar in general appearance to *A. squamata*, but has longer arms and longer arm spines, and its oral shield is slightly longer than its width. See the table on page 102. Because of the difficulties in distinguishing *A. squamata* from *A. pugetana*, some records are suspect.

Distribution

Broadly cosmopolitan in boreal, temperate and tropical waters, but not found in polar regions; upper intertidal to 1330 metres. RBCM specimens 0–133 metres; Austin collection 0–37 metres.

Biology

Habitat: *Amphipholis squamata* does not burrow. It is abundant in algal holdfasts, under rocks and in the shell-gravel of shallow intertidal pools. Its tolerance of low oxygen levels may allow this species to survive in tidal pools where algae deplete the oxygen at night.

Feeding: *A. squamata* feeds on suspended material and bottom material, such as unicellular algae, protozoans, small animals and detritus.

Locomotion: Rowing movements of the arms propel this brittle star rapidly over the bottom. It can also climb vertical surfaces and when dislodged curl up in a ball to fall to the bottom.

Reproduction: This species is a simultaneous hermaphrodite (possessing both eggs and sperm at the same time) although the testes may mature before the ovaries. Throughout the year only one egg matures at a time in each of the ten ovaries. Each testis produces a pulse of sperm but the pulses are not synchronized with testes in other arms. Sperm and one egg are released into the genital bursa where fertilization occurs. The parent broods the embryos to an advanced juvenile stage, capable of crawling from the genital bursae. The release of juveniles seems to peak in winter and in summer. With no swimming larva there are often dense populations in a confined area.

Other Notes: Its association with floating materials, and its ability for self-fertilization, may account for the widespread distribution of this species. When disturbed, *A. squamata* emits light (bioluminescence) from the spinal ganglia. The black and spotted varieties and brooding parents produce more light than other individuals. The light production is thought to startle predators or, if the arm breaks off, to distract them. The orthonectid *Rhopalura ophiocomae* parasitizes the ovary and prevents reproduction in 24–50 per cent of the population on San Juan Island, Washington.

References

Due to the wide distribution of this species there are many references, and this is a selected list: Austin and Hadfield 1980; Boolootian and Leighton 1966; Brehm and Morin 1977; Deheyn, Mallefet and Jangoux 1997; Emson and Foote 1979; Emson and Wilkie 1982; Hendler 1975, 1996b; Rader 1982; Strathmann and Rumrill 1987. **Range:** Hendler 1996b.

Family Ophiotrichidae

Characterized by the presence of a dense group of dental papillae coupled with the absence of oral papillae. The dorsal side of the disc extends over the top of the arms; long, erect arm spines. Primarily a tropical family, and only one species has been recorded in British Columbia.

Ophiothrix spiculata   **Glass-spined Brittle Star**

Description
Disc: Up to 18 mm in diameter, the dorsal surface covered by prominent erect spines carrying rows of thorn-like spinelets; the spines continue down the ventral sides between the arms.

Arms: Five to eight times as long as the disc diameter, with up to eight erect, glassy arm spines bearing a row of spinelets along each edge. The most ventral spines near the arm tips are hooked. The arms have no supplementary dorsal plates.

Mouth: The apex of each jaw has a pad of dental papillae; there are no oral papillae. There is hole between the adjacent jaw plates proximal to the oral shield. **See colour photo C-38.**

Colour: Varies widely from red, orange or yellow to black, offset by white arm spines. **See colour photo C-37.**

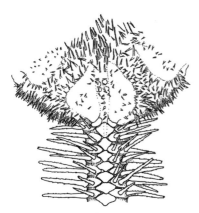

59. *O. spiculata* dorsal.

60. *O. spiculata* ventral.

Taxonomic Notes: Includes *Ophiothrix dumosa* Lyman, 1860. The name *spiculata* is from the Latin *speculum*, meaning "point" or "spike".

Similar Species

Ophiothrix spiculata superficially resembles *Ophiopholis aculeata*, but the latter has oral papillae and no dental papillae. *O. aculeata* adults do not have a hole between the jaw plates or arm spines with lateral spinelets, though some juveniles have spines with spinelets.

Distribution

Published records are from northern Chile and the Galapagos Islands to Moss Beach, California. But Val McDonald of Biologica Environmental Services has identified two juveniles from the waterfront in Victoria, BC, which we confirmed as *Ophiothrix spiculata*. Other specimens from Canada previously thought to be *O. spiculata* have been reidentified as *Ophiopholis* spp. The specimens from Victoria may have been introduced via shipping or they may reflect a disjunct distribution of the species, as suggested for *Ophiopteris papillosa* and *Ophioplocus esmarki*. This species ranges from the low intertidal zone to 2059 metres depth, but is typically in shallow subtidal waters.

Biology

Habitat: *Ophiothrix spiculata* lives under rocks and in rock crevices and kelp holdfasts where it can temporarily attach by one or more arms. In southern California, populations of 80 per 0.1 square meter have been recorded.

Feeding: While suspension feeding, this species extends several arms into moving water. The tube feet are extended and may be oriented at right angles to the current or alternatively in a cross-forming array down the arm. Particles adhere to the sticky spines and feet. The feet curl around the spines and other feet to "lick off" the particles. Then the pairs of tube feet pass a bolus of particles down the arm to the jaws, which close several times as the dental papillae compact the bolus before it is pushed into the gut. *O. spiculata* also feeds on carrion and detritus, transferring the food to the mouth by rolling arm loops.

Reproduction: Spawning has been observed in July in central California. Both eggs and sperm are broadcast and the floating embryos develop into free-swimming ophiopluteus larvae.

Other Notes: In California, the sea star *Astrometis sertulifera* and several fishes, such as Rock Wrasse (*Halichoeres semicinctus*), Pile Perch (*Rhacochilus vacca*) and Sand Bass (*Paralabrax nebulifer*) prey on Glass-spined Brittle Stars.

References
Austin and Hadfield 1980; McClendon 1909; Ricketts, Calvin and Hedgpeth 1985; Rumrill 1982; Rumrill and Pearse 1985.
Range: Austin and Hadfield 1980; Austin and Haylock 1973.

Family Ophiactidae

Most species in this family have conspicuous radial shields. All have one apical oral papilla separated by a space from the lateral oral papillae, and none have dental papillae. The arm spines are short, pointed and erect (i.e., not pressed against the side of the arm). Five species in one genus, *Ophiopholis*, live in the eastern North Pacific.

Distinguishing characters for species in *Ophiopholis*					
Character	*O. aculeata*	*O. kennerlyi*	*O. japonica*	*O. bakeri*	*O. longispina*
Radial shield covering	Some spines or granules.	Granules.	Naked or a few long spines.	Abundant, furry.	Naked.
Disc covering	Granules or thick spines.	Granules.	Granules and often spines.	Abundant short spines.	Long, thin spines.
Longest arm-spine length	0.75–1.5 arm joints.	0.5 arm joint.	1–2 arm joints.	2 arm joints.	3 arm joints.
Swollen dorsal arm plates	No.	No.	No.	No.	Yes.
Position of supplementary arm plates	Contiguous.	Contiguous.	Contiguous.	Separated.	Separated.
Number, shape of supplementary arm plates	10–24, angular.	6–10, angular.	14 or more, angular or round.	14–20, round.	About 8, round.
Shape of arm spine	Flat, blunt.	Flat, blunt.	Conical, pointed.	Conical, pointed.	Conical, pointed.
Depth in this region	Intertidal to 366 m.	Intertidal to 435 m.	15–1884 m.	18–1204 m.	507–1253 m.

Ophiopholis

Species in this genus have granules or spinelets covering their disc scales, except for the primary plates. Small supplementary plates surround the dorsal arm plates. The most ventral arm spine is often modified into a hook, especially at the distal end of the arm.

The type species of the genus is *Ophiopholis aculeata*, first described by Linnaeus in 1767. Clark (1911) noted that it occurred in the Atlantic, Arctic and Bering Sea; but he considered two other species described earlier by Lyman in the northern Pacific, *O. kennerlyi* and *O. japonica*, to be varieties of *O. aculeata*. Now, experts agree with Lyman and have raised the two varieties to species status. The typical form of *O. aculeata* (variety *typica* in some references) occurs at least into British Columbia, but has not been positively identified south, in the waters of Oregon and California. Its true distribution is still not clear, because specimens of uncertain identification in this region have been listed in the literature simply as *O. aculeata* with no variety indicated. Thus, in the region covered by this book we include five species, *O. aculeata*, *O. kennerlyi*, *O. japonica* and two species from deeper water, *O. longispina* and *O. bakeri*.

Characteristics of the five species in this region are shown in the table on the facing page.

Ophiopholis aculeata

Description
Disc: Up to 20 mm in diameter, and the plates bear granules, often with spinelets scattered among them. Circular primary plates are usually visible on the disc. Radial shields are large and triangular but concealed. The ventral side may not have granules or spinelets.
Arms: About four times longer than the diameter of the disc. The dorsal arm plates are transverse-oval, surrounded by 10 to 24 small, angular supplementary plates in contact with each other. The ventral arm plates are rectangular and separated by small depressions. Each segment has six to seven short, stout arm spines, from 0.75 to 1.5 arm joints long; the lowest spine is modified into a hook with a smaller hook below the tip, especially distally. Each podial pore has one large tentacle scale.

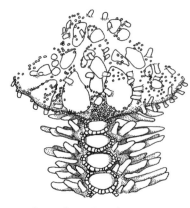

61. *O. aculeata* dorsal.

Mouth: Each side of the jaw has one small apical papilla and three or (rarely) two oral papillae. The oral shields are oval, but variable.

Colour: Highly variable, mostly red, often variegated. The arms often have bands of colour.

Taxonomic Notes: *Ophiopholis aculeata* was described by Linnaeus in 1767, from European seas. In some older literature it may be referred to as *O. aculeata* variety *typica* to differentiate it from the other forms found in the North Pacific, which have since been raised to full species status. The species name comes from the Latin *aculeatus*, meaning "prickly", referring to the spinelets on the disc.

Similar Species
See the table on page 110.

Distribution
May be circumpolar, but there are no records over much of the Canadian and Russian Arctic. Along the Scandinavian coast as far as the south coast of Britain. On the North American side of the Atlantic south to Cape Cod. In the Pacific in the Bering Sea, and south into Japan in the west and into southeast Alaska and British Columbia in the east. Usually shallow subtidal, rarely below 300 metres, but has been recorded at 1000 metres in the Faroe Channel.

Biology
Habitat: *Ophiopholis aculeata* is usually found nestled in hollows and crevices of stones and shells.

Feeding: Its food appears to be mainly detritus, but it has been observed occasionally using the arm-loop capture method.

Reproduction: In the Atlantic the eggs are numerous and small. When full of mature eggs or sperm in June and July the interradial areas of the disc bulge out between the arms. During spawning it raises the disc off the substrate and pushes other brittle stars away.

Other Notes: Eaten by cod.

References

Austin and Hadfield (1980), Ivanov (1964), Litvinova (1981), Lyman (1879), Mortensen (1977), Serafy (1971). **Range:** Austin and Haylock (1973), Clark (1911), D'yakonov (1967), Grainger (1955).

Ophiopholis bakeri

Description

Disc: Round to pentagonal, about 10 mm diameter, with inflated interradial regions. Pointed granules and spinules densely cover the thin, overlapping scales and radial shields. Spinules have two to five microscopic points. The radial shields are twice as long as their width; they touch distally but not proximally.

Arms: Four to six times as long as the disc diameter. The dorsal arm plates are rounded proximally but more oval and elongated distally. Supplementary plates around the dorsal arm plates are separated and granular in shape with one or two points. Each arm has up to seven spines, each two to threee arm joints in length; the dorsal spine is usually the longest, and the lowest spine is small with two glassy hooks at the tip. The ventral arm plates are octagonal, with the distal edge concave and the proximal convex. The first ventral arm plate is small and enclosed by adoral shields;

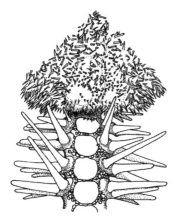

62. *O. bakeri* dorsal.

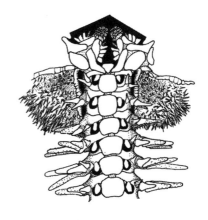

63. *O. bakeri* ventral.

the next few proximal plates have a shallow groove down the middle. Each podial pore has a single flat, elongated tentacle scale with an expanded tip.

Mouth: Each side has four to six flat-to-spine-like oral papillae. The oral shield is diamond shaped and twice as wide as its length. The adoral shields are nearly as large as the oral shields, but L-shaped with the distal ends touching the second ventral arm plate.

Colour: Reddish-brown disc, with bluish-grey in the centre; the arms and spines are pinkish dorsally and yellowish-white with brown spots ventrally. **See colour photo C-34.**

Taxonomic Note: The name *bakeri* is after Frederick Baker, 1854–1938, a malacologist and collector from San Diego, California.

Similar Species

Ophiopholis kennerlyi and *O. longispina* are similar to *O. bakeri*. See the table on page 110.

Distribution

Southeast Alaska, near Sitka, to Baja California; 18–1204 metres. One record from near Kodiak Island needs confirmation. RBCM collection 18–315 metres; Austin collection 37–1204 metres.

Biology

Habitat: *Ophiopholis bakeri* inhabits a variety of hard substrates. While typically few in number, it is often abundant on the surface of the Boot Sponge (*Rhabdocalyptus dawsoni*). Here it is likely protected from predators by the net of spicules projecting from the sponge.

Feeding: *O. bakeri* is similar in external structure to *O. kennerlyi*, a suspension feeder that attaches to sessile animals or other raised substrates bathed by currents, so it likely feeds in the same way. *O. bakeri* may take advantage of the currents generated by the Cloud Sponge (*Aphrocallistes vastus*) to obtain food.

References

Boolootian and Leighton 1966; Clark 1911; Hendler 1996b; McClendon 1909. **Range:** Astrahantseff and Alton 1965; Austin and Haylock 1973; Kyte 1969.

Ophiopholis japonica

Description
Disc: Up to 22 mm in diameter, covered by granules and usually some spines. The radial shields are exposed.

Arms: The oval dorsal arm plates are surrounded by 14 or more contiguous supplementary plates. The arm spines are conical and pointed; the longest is two to three arm joints long.

Mouth: One small apical oral papilla, and two or three large, rounded oral papillae on each side of jaw.

Colour: Reddish-brown in the centre of the dorsal side of the disc; pinkish-white on the arms.

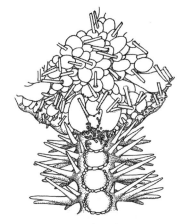

64. *O. japonica* dorsal.

See colour photo C-33.

Taxonomic Note: Initially described as a species by Lyman in 1879. Later, H.L. Clark concluded that *Ophiopholis japonica* is a variety of *O. aculeata*, based on specimens that appeared close to *O. kennerlyi*, while others were close to *O. bakeri* and *O. longispina*. Imaoka et al. consider *O. japonica* to be a separate species. The name *japonica* is after Japan, where the species was first collected.

Similar Species
See the table on page 110.

Distribution
Aleutian Islands to Knight Inlet, British Columbia, and the Bering Sea to Japan at latitude 33°N; 15–1884 metres. RBCM collection 80–640 metres; Austin specimens 90–1204 metres.

Biology
Habitat: *Ophiopholis japonica* lives on a wide range of substrates, including broken shells, gravel, sand and mud.

Feeding: Likely similar to *O. kennerlyi* (see page 116).

References
Range: Austin and Haylock 1973; Clark 1911; D'yakonov 1967; Imaoka, Irimura, Okutani, Oguro, Oji, Shigei and Horikawa 1990.

Ophiopholis kennerlyi   Daisy Brittle Star

Description
Disc: Up to 20 mm in diameter. Round plates bear short, small granules on the dorsal side, tending to spinelets on the side and oral part. The radial shields are large and triangular, but concealed by granules. The disc is often inflated in the interradial regions between the arms. Circular primary plates are usually visible.

Arms: Short and thick, capable of bending upward. Transverse-oval dorsal arm plates are surrounded by small, supplementary plates in contact with each other. The ventral arm plates are rectangular and separated by small depressions. Each segment has five to seven short, stout, slightly compressed arm spines, the lowest modified into a hook with a smaller hook below the tip, especially distally; each podial pore has one large tentacle scale.

Mouth: One small apical oral papilla, and two or three large, rounded oral papillae on each side of jaw.

Colour: Highly variable in red, purple, white, plain or patterns, with rarely two specimens alike. Arms often have bands of colour. **See colour photos C-39 and C-40.**

Taxonomic Notes: Initially described as a species by Lyman in 1860. H.L. Clark (1911), studying a very large collection of related forms extending from California to Japan, concluded that *Ophiopholis kennerlyi* is a variety of *O. aculeata*. When reviewing the manuscript for this book, Gordon Hendler stated that he is treating *O. kennerlyi* as a valid species in the revised *Light and Smith Manual* (in press), and we have followed him here. The general distributions differ between *kennerlyi, japonica* and *aculeata*, but there is some overlap in the Bering Sea. *O. caryi* is a synonym of *O. kennerlyi*. This species was probably named after Dr Caleb B.R. Kennerly (1829–61), surgeon and naturalist on several expeditions, including the Northwest Boundary Survey (1857–61).

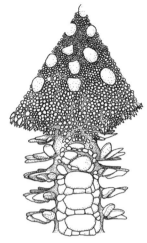

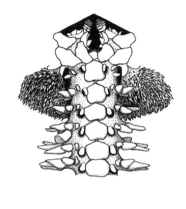

65. *O. kennerlyi* dorsal.　　　**66**. *O. kennerlyi* ventral.

Similar Species
See the table on page 110.

Distribution
From central Alaska to Santa Barbara, California; intertidal to 435 metres.

Biology
Habitat: This species is usually found in areas with rock or shell bathed by tidal currents. It hides its disc in small rubble, among bryozoan or hydrocoral colonies or other cover, and exposes just the arms when feeding.

Feeding: *Ophiopholis kennerlyi* extends its arms up into the water and captures suspended particles on the mucus-covered tube feet which may extend out from or cross over the arm. The tube feet also pick up food from the substrate. One or more tube feet bend toward the disc and curl around the next set of tube feet transferring the bolus of food and mucus to the mouth. The arms can rapidly coil around larger food items, and then transfer them directly to the mouth.

Reproduction: The sexes are separate. Males produce sperm all year round, and spawning has been recorded in the laboratory for most of the year in the San Juan Islands. The orange oocytes (eggs), 100–105 μm in diameter, develop into an ophiopluteus larva. When conditions are good for plankton, the ophiopluteus larva has been shown to produce clones of itself that can take advantage of the situation and potentially increase the geographic range and number of juveniles without additional effort by the parent.

References
Austin and Hadfield 1980; Balch, Hatcher and Scheibling 1999; Balser 1998; Fontaine and Lambert 1976; Hendler 1975, 2007; Labarbera 1978; Scouras, Beckenbach, Arndt and Smith 2004; Strathmann and Rumrill 1987; Summers, Hylander, Colwin and Colwin 1975. **Range:** Austin and Hadfield 1980; Austin and Haylock 1973; Clark 1911; Kyte 1969.

Ophiopholis longispina

Description
Disc: Up to 10 mm in diameter, covered mostly by large radial shields that are longer than their width and just touching distally. Small circular or oval plates occupy spaces between the shields and are often armed with a single slender spine that may be smooth or rough and have several teeth at the tip. The ventral interradial areas are covered by spine-bearing plates.
Arms: The dorsal arm plates are circular and swollen and surrounded by about eight round, supplementary granules that are not in close contact. Each segment has six to eight slender, bluntly pointed arm spines, the uppermost about three arm joints long; each podial pore has a single blunt tentacle scale.
Mouth: The oral shield is short and wide, similar to that of *Ophiopholis bakeri*. The oral papillae are short, flat and blunt, and clustered near the adoral shield. The first ventral arm plate is tiny; subsequent plates are hexagonal, about as long as their width.
Colour: Bluish-white disc, red arm plates and beige arm spines. **See colour photo C-35.**
Taxonomic Note: The name *longispina* is from the Latin *longus*, meaning "long", and *spina*, meaning "thorn", referring to the long arm spines.

Similar Species
See the table on page 110 and compare with *Ophiopholis kennerlyi* and *O. bakeri*.

Distribution
Queen Charlotte Islands to San Diego, California; 507–1253 metres. RBCM records 591–1259 metres; Austin collection 924–1061 metres.

Biology
Habitat: Often collected in association with sponges.

References
Clark 1911. **Range:** Astrahantseff and Alton 1965; Austin and Haylock 1973; Clark 1911; Kyte 1969.

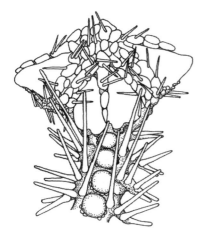

67. *O. longispina* dorsal.

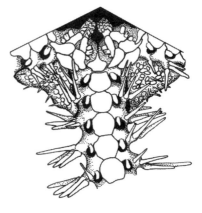

68. *O. longispina* ventral.

Family Ophiocomidae

Disc covered by fine granules or naked skin, sometimes bearing scattered spines; radial shields stout but covered by skin; mouth has four to six oral papillae on each side, and the dental papillae are well developed, forming a clump at the apex of each jaw; arm spines long, stout and perpendicular to the arm axis; podial pores have one or two tentacle scales.

Ophiopteris papillosa

Description
Disc: Up to 13 mm in diameter, covered by stout cylindrical stumps. The ventral interradial area is covered by more slender stumps.

Arms: Overall appearance robust and spiny, 3 to 4.5 times as long as the disc diameter and, including spines, as wide as the disc diameter. The dorsal arm plates are hexagonal and broader than their length; each side has five flat, blunt arm spines. The dorsal arm spine is very short; the most ventral and sometimes the adjacent spine are reduced to scales; and the middle and sometimes the fourth are as long as three arm joints; all are finely serrated toward the tips.

Mouth: Each side has four to six pairs of small oral papillae, with

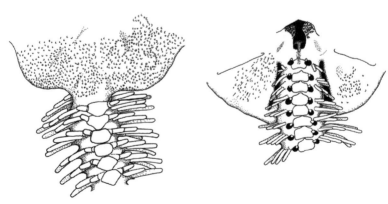

69. *O. papillosa* dorsal.   **70**. *O. papillosa* ventral.

one pair in the corner of the mouth (actually tentacle scales). The apex of each jaw has a well-developed clump of dental papillae. The mouth plates are almost triangular, with rounded corners. The adoral shields are narrow and slightly flared distally. **See colour photo C-30.**

Colour: Individuals in the intertidal zone are deep chocolate brown with bands on the arms, and those in subtidal areas are a lighter tan colour. **See colour photo C-31.**

Taxonomic Notes: Originally *Ophiocoma papillosa* Lyman; revised to *Ophiopteris papillosa* by Clark (1911). The name *papillosa* is from the Latin *papilla*, meaning "nipple" or "bud", referring to the disc covered with cylindrical stumps.

Similar Species
Different from any other species in the region.

Distribution
Barkley Sound, British Columbia, and from southern Oregon to Isla Cedros, Baja California; low intertidal to 140 metres. North of southern Oregon the species has only been found in three localities near Bamfield, BC, in 10 to 12 metres (Austin Collection). The hiatus of records in Washington and central and northern Oregon suggests that the species has a disjunct distribution.

Biology
Habitat: In California it is fairly common under rocks and in algal holdfasts and sometimes associated with the Purple Sea Urchin (*Strongylocentrotus purpuratus*).

Feeding: *Ophiopteris papillosa* can feed with its arms raised up off the substrate and tube feet extended out laterally in two regular rows to form a double comb-like net. Particles adhere to mucus secretions and are passed down to the mouth. This species captures larger organisms and carrion by rapidly coiling its arms around the food; it even breaks up food held in the mouth by using the arm spines as levers. Captive specimens have climbed the walls of their aquarium and, apparently, fed off the surface film.

Reproduction: In California the reproductive period is from early December to mid June. Gonad growth begins in August and September and spawning occurs in January, when individuals can produce more than 100,000 eggs. The egg diameter of 102 μm suggests that this species produces feeding larvae.

References
Austin and Hadfield 1980; McClendon 1909; Ricketts and Calvin 1964; Rumrill 1982; Rumrill and Pearse 1985. **Range:** Austin 2000; Austin and Hadfield 1980; Austin and Haylock 1973.

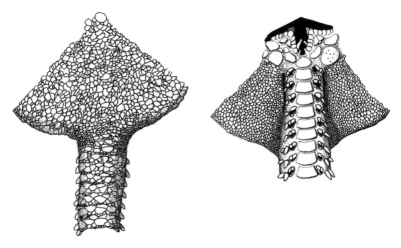

71. *O. esmarki* dorsal.   **72**. *O. esmarki* ventral.

Family Ophiolepididae

Disc covered with thin scales or plates; radial shields usually stout; mouth has a single series of oral papillae and the second oral tentacle pore opens within the oral slit; genital scales not leaf-like; short, stout arms merge into the disc; stout arm spines lie flat on the sides of the arms.

Ophioplocus esmarki

Description
Disc: Up to 30 mm in diameter, covered with overlapping scales, thick and irregularly shaped, like pebbles; about 12 larger rounded, swollen scales line the edge of the disc. The disc scales continue onto the arms. The radial shields are small (about 2 mm) and similar to the surrounding scales.

Arms: Short and stubby, three times as long as the disc diameter, with three blunt arm spines, almost as long as a joint, at acute rather than right angles. The dorsal arm plates are small, angular pieces with smaller supplementary plates between them. The ventral arm plates are broad and pentagonal, 1.5 times wider than their length. Each pore has two flat, oval tentacle scales, side by side.

Mouth: Each side has five or six stout, crowded, oral papillae. The oral shields are heart shaped and the adoral shields are elongated, swollen triangles.

Colour: Brown or red-brown on the dorsal surface, lighter on the ventral surface and arm spines. **See colour photo C-36.**

Taxonomic Note: The name *esmarki* is after Laurits M. Esmark, 1806–84, a zoologist from Norway. Lyman refers to Professor Esmark collecting numerous specimens near San Diego, California.

Similar Species
At first glance, *Ophioplocus esmarki* might be confused with a stubby armed *Ophiura sarsii*, but it is smaller, and lacks obvious radial shields and the split dorsal arm plates with supplementary plates.

Distribution
Quatsino Sound, BC, to San Diego, California; intertidal to 70 metres. The Quatsino Sound record (RBCM) is the only one north

of central California. Lambert collected it in a patch of sand-shell among rocks and boulders at 15 metres depth in an exposed bay. Because the species is ovoviviparous (see Reproduction, below), one individual transported north on something like a floating kelp holdfast could have established a colony.

Biology
Habitat: In California, *Ophioplocus esmarki* occurs in intertidal pools in sandy mud under flat rocks, and in kelp holdfasts and crevices.

Feeding: Slow moving, this brittle star eats small, slow-moving animals or carrion. It moves food particles to its mouth via adhesive tube feet; the arms can bend in an inverted "U" around food to bring more tube feet into play.

Reproduction: This ovoviviparous species broods eggs in its bursa, where they develop directly into small brittle stars that escape via the genital slits. In California, the bursae are swollen with eggs in January, and the size of the brood is highest at the beginning of the brooding period in April and May. The brooding period varies from year to year. In Monterey, young occur in brood pouches in July. Animals with a disc diameter of about 18 mm can produce 2,000 eggs with a diameter of 328 µm.

References
Austin 2000; Austin and Hadfield 1980; MacGinitie and MacGinitie 1968; McClendon 1909; Ricketts and Calvin 1964; Rumrill 1982; Rumrill and Pearse 1985. **Range:** Austin 2000; Austin and Hadfield 1980.

GLOSSARY

aboral Opposite or away from the mouth. In most echinoderm groups it refers to the dorsal surface; but in crinoids, which have their mouth on the upper side, aboral refers to the ventral surface.

adoral Toward or in a position closer to the mouth.

adoral shield One of two plates in ophiuroids adjacent to an oral shield closer to the mouth. See figure 25.

ambulacra One of the regions of the body associated with the tube feet of the water-vascular system. Most echinoderms have five ambulacral regions.

ampulla A sac-like structure at the inner end of a tube foot that contracts to extend the foot by hydraulic pressure. See figure 10.

apical oral papilla The papilla (alone or in a pair) at the apex of the jaw and ventral to the teeth of ophiuroids. See figure 25.

apical system A set of plates at the apex of a regular sea urchin's test that consists of five genital and five ocular plates surrounding a flexible plated membrane that contains the anus. See figure 14.

arm Appendage extending radially and laterally from the disc of ophiuroids or asteroids.

arm comb A series of spinelets, typically in a comb-like row. The primary arm comb is at the edge of the disc just above each arm of ophiuroids, and the secondary arm comb is just below. See figures 25 and 40.

arm length In ophiuroids, measured from the edge of the disc to the tip of an unbroken arm. In asteroids, usually measured from the centre of the disc to the tip of an arm.

arm spine In ophiuroids, one of a series of spines on the lateral arm plate. Ventral arm spines may be difficult to distinguish from tentacle scales.

benthic Of or at the ocean bottom.

brachial   In brittle stars, the arm. In feather stars, one of the many calcareous pieces that make up the arms (see figure 5), connected by sets of ligaments, flexor muscles or elastic fibres.

bursa   One of a pair of pouches in the disc of ophiuroids adjacent to the base of each arm. Gonads usually empty their eggs or sperm into the bursa before release. Brooding species retain their larvae in the bursa.

calyx   Ossicles fused together in a cup-shape that contains or supports the soft parts (organs) of a crinoid.

cirri (singular: **cirrus**)   Calcareous appendages at the base of a feather star that grasp the substrate; made up of multiple joints with a hook at the tip. See figures 3 and 7.

dental papilla   One of a cluster of spinelets on the apex of each jaw just ventral to the teeth. Characteristic of the families Ophiotrichidae and Ophiocomidae. See figure 60.

disc   Central part of the body to which the arms attach.

disc diameter   In ophiuroids, measured from the disc edge above an arm to the opposite edge between a pair of arms. The measurement varies depending on material in the gut or gonads.

disc notch   An indentation of the disc above each arm of a brittle star. See figure 25.

distal   Away from the centre of the disc or central axis.

dorsal   The back or top side of an animal. In almost all echinoderms, this is the opposite side that the mouth is located on (see **aboral**), but crinoids have the mouth on the top side.

echinopluteus   A echinoid larval type with long arms supported by rods of calcite. See figure 16.

genital papillae   A series of small ossicles along the edge of the genital slit of brittle stars; sometimes continuous with a dorsal arm comb. See figures 25 and 41.

genital plate   In brittle stars, an elongated ossicle along the edge of a genital slit (see figure 25). In sea urchins, a plate (see figure 14) at the top end of the interambulacral column that alternates with the ocular plates.

genital slit   A slit or pore on the oral side of a brittle star disc adjacent and parallel to each arm, into which the gonad empties; also called the bursal slit. See figure 25.

gonoduct   Tube that connects the male or female gonad to the outside via the gonopore.

gonopore   The external opening through which eggs or sperm are released.

granule   A minute, more-or-less spherical ossicle.

interambulacra   In echinoids, two rows of plates in the test between the rows that are perforated for the passage of the tube feet. See figure 9.

interradial   The sector of the disc between two arms of a brittle or sea star (in brittle stars, also called interbrachial).

jaw A pair of ossicles forming each of five triangular structures around the mouth in brittle stars. In sea urchins the jaw is a complicated structure called Aristotle's lantern (see figure 10 and page 24).

jaw plate One of a pair of large ossicles forming each jaw in brittle stars (also called an oral plate). See figure 25.

lateral arm plate A curved plate covering the lateral surface of each arm segment of a brittle star and typically bearing arm spines. See figure 25.

madreporite A modified oral shield with one or more pores connecting to the water-vascular system. See figure 14.

mamelon A knob-like tubercle on which a moveable spine articulates. See figure 11.

mesentery A thin sheet of tissue that attaches various organs to the body wall.

miliary spines In echinoids, small spines that are shorter than the regular spines and form an undergrowth below the main canopy. "Miliary" refers to something the size of a millet seed.

mouth In brittle stars and sea urchins, an opening on the ventral side leading to the gut and guarded by the jaws. In feather stars, an opening into the gut where the ambulacral grooves meet on the dorsal side of the central tegmen. See figures 3 and 10.

ocular plate A small plate at the dorsal end of the ambulacral series in sea urchins, bearing an ocular tentacle pore and situated between two genital plates. See figure 14.

ophiopluteus A larval type of many brittle stars where the arms are supported by skeletal rods and bear bands of cilia.

oral papilla One of a series of small ossicles along the edge of a jaw, or attached to a jaw plate or an adoral shield of brittle stars. See figure 25.

oral shield A large shield or plate in each ventral interradius distal to the jaws. See figure 25.

oral tentacle One or two of the most basal tube feet that emerge between the jaws of brittle stars.

ossicle A single calcified element of an echinoderm skeleton.

papilla In brittle stars, a small projecting ossicle. In feather stars, usually a small elevation of soft tissue bearing the mouth or opening of the gonad.

pedicellariae (singular: **pedicellaria**) Small pincer-like structures that occur in numbers on the outer surface of many echinoids. These pincers consist of a muscular stem stiffened by a calcareous rod, and a moveable head, usually made up of three jaw-like valves. See figure 13.

pentacrinoid stage Juvenile stage of a feather star that has a stalk and five arms. With development, the stalk is replaced with cirri and the arms divide and multiply. See page 10.

pinnule In feather stars, each arm joint (brachial) has a side branch called a pinnule with successive pinnules on alternate sides of the arm.

See figures 3, 6 and 8.

pluteus (plural: **plutei**)   The long armed, V-shaped larva of an ophiuroid (see ophiopluteus) and echinoid (see echinopluteus).

podial pore   A hole in a calcareous plate of an echinoid or ophiuroid through which a tube foot passes. See figures 17, 19 and 21.

primary disc plates   A set of mid-dorsal and five adjacent plates on the radii of a brittle star disc often forming an obvious rosette pattern. See figure 25.

proximal   Toward the centre of the disc or central axis.

radial   Radiating from the centre of the disc; aligned with the arms.

radial shield   One of a pair of large plates near the edge of a brittle star disc where the arm attaches. See figure 25.

RBCM   Royal British Columbia Museum.

ring canal   Part of the water-vascular system that surrounds the mouth. See figure 10.

spicules   Structures made of silica that form the skeleton of glass sponges.

spinelet   A small spine-like ossicle.

spine   A long ossicle that may be rounded or square ended as well as pointed. A spine typically has a muscular articulation with another ossicle.

substrate   The surface or material on which an organism lives.

supplementary arm plates   Smaller plates adjacent to the dorsal, lateral and ventral arm plates of brittle stars. See figure 65.

syzygy   A rigid joint between two brachials in a crinoid arm, holding the brachials together with ligaments rather than muscles. The external suture looks like a perforated line – its position is an important diagnostic feature. See figures 5 and 6.

tegmen   The soft tissue of the oral surface around the mouth. It may be naked, have small calcareous pieces, or be covered with small plates or nodules. The tegmen bears the anus, at the apex of a small tube or papilla, and ambulacral grooves connected to the mouth. See figure 4.

tentacle   In brittle stars, a tube foot.

tentacle pore   An opening through which a tube foot projects.

tentacle scale   An ossicle at the edge of a tentacle pore along the arm of a brittle star. Typically much smaller than an arm spine, but sometimes not clearly different in form. Ossicles associated with the oral tentacle pores may be called oral papillae in some groups. See figure 25.

test   The spherical skeleton of a sea urchin made up of many small plates that bear the spines and pedicellariae. See figure 9.

tooth   In sea urchins, one of five calcareous structures in the mouth for tearing up seaweed. In brittle stars, one of a series of ossicles on the proximal edge of the jaw, dorsal to any oral papillae.

tube foot   A hollow, tubular structure that is an extension of the water-vascular system of echinoderms and aids in locomotion or food gathering. In brittle stars, a tube foot is called a tentacle and it extends from a tentacle pore on each side of the ventral arm plate. In feather stars, tube feet line the ambulacral groove on each arm. In sea urchins, they protrude through the ambulacral pore pairs and usually have a sucker on the end.

water-vascular system   An internal system of liquid-filled canals and vesicles that uses hydraulic pressure and muscular contraction to operate the tube feet.

ventral   Usually the bottom side of an animal in its normal orientation. In sea urchins and brittle stars, it is the oral side, but in feather stars it is aboral.

ventral arm plate   A single plate covering the ventral surface of each arm segment of a brittle star. See figure 25.

vertebra (plural: **vertebrae**)   In brittle stars, a segment of the arm formed by the fusion of a pair of ambulacral ossicles and articulating with vertebrae of adjacent arm segments; usually covered by the arm plates.

REFERENCES

Astrahantseff, S., and M.S. Alton. 1965. Bathymetric distribution of brittle stars (Ophiuroidea) collected off the northern Oregon coast. *Journal of the Fisheries Research Board of Canada* 22: 1407–24.

Austin, W.C. 1966. Feeding mechanisms, digestive tracts and circulatory systems in the ophiuroids *Ophiothrix spiculata* Le Conte, 1851, and *Ophiura luetkeni* (Lyman, 1860). Ph.D. thesis, Stanford University, California.

———. 1985. *An Annotated Checklist of Marine Invertebrates in the Cold Temperate Northeast Pacific*. Cowichan Bay, BC: Khoyatan Marine Laboratory.

———. 2000. Rare and endangered marine invertebates in British Columbia. In *The Biology and Management of Species and Habitats at Risk*, Edited by L.M. Darling. Victoria: British Columbia Ministry of Environment, Lands & Parks; and Kamloops, BC: University College of the Cariboo.

Austin, W.C., and M.G. Hadfield. 1980. Ophiuroidea: The brittle stars. In *Intertidal Invertebrates of California*, edited by R.H. Morris, D.P. Abbott and E.C. Haderlie. Stanford, California: Stanford University Press.

Austin, W.C., and M.P. Haylock. 1973. *BC Marine Faunistic Survey Report: Ophiuroids from the Northeast Pacific*. Technical Report 426. Ottawa: Fisheries Research Board of Canada.

Balch, T., B.G. Hatcher and R.E. Scheibling. 1999. A major settlement event associated with minor meteorologic and oceanographic fluctuations. *Canadian Journal of Zoology* 77: 1657–62.

Balser, E.J. 1998. Cloning by ophiuroid echinoderm larvae. *Biological Bulletin* (Woods Hole, Massachusetts) 194: 187–93.

Baranova, Z.I. 1966. Class Echinoidea. In *Atlas of the Invertebrates of the Far Eastern Seas of the USSR*, edited by E.N. Pavlovskii. Jerusalem: Israel Program for Scientific Translations.

Bazhin, A.G. 1998. The sea urchin genus *Strongylocentrotus* in the seas of Russia: taxonomy and ranges. In *Echinoderms: San Francisco – Proceedings of the Ninth International Echinoderm Conference, San Francisco, California 5-9 August 1996*, edited by R. Mooi and M. Telford. Rotterdam: A.A. Balkema.

Berger, J., and R.J. Profant. 1961. The ectocommensal ciliate fauna of the Pink Sea Urchin, *Allocentrotus fragilis*. *Journal of Parasitology* 47: 417–18.

Bernard, F.R. 1977. Fishery and reproductive cycle of the Red Sea Urchin, *Strongylocentrotus franciscanus*, in British Columbia. *Journal of the Fisheries Research Board of Canada* 34: 604–10.

Biermann, C.H. 1998. Population genetic structure and the evolution of reproductive isolation in strongylocentrotid sea urchins. In *Echinoderms: San Francisco – Proceedings of the Ninth International Echinoderm Conference, San Francisco, California 5-9 August 1996*, edited by R. Mooi and M. Telford. Rotterdam: A.A. Balkema.

Biermann, C.H., B.D. Kessing and S.R. Palumbi. 2003. Phylogeny and development of marine model species: strongylocentrotid sea urchins. *Evolution and Development* 5: 360–71.

Birenheide, R., and T. Motokawa. 1998. Crinoid ligaments: Catch and contractility. In *Echinoderms: San Francisco – Proceedings of the Ninth International Echinoderm Conference, San Francisco, California 5-9 August 1996*, edited by R. Mooi and M. Telford. Rotterdam: A.A. Balkema.

Birkeland, C., and F.S. Chia. 1971. Recruitment risk, growth, age and predation in two populations of sand dollars, *Dendraster excentricus* (Eschscholtz). *Journal of Experimental Marine Biology and Ecology* 6: 265–78.

Bluhm, B.A., D. Piepenburg and K. Von Juterzenka. 1998. Distribution, standing stock, growth, mortality and production of *Strongylocentrotus pallidus* (Echinodermata: Echinoidea) in the northern Barents Sea. *Polar Biology* 20: 325–34.

Boolootian, R.A. 1966. *Physiology of Echinodermata*. New York: Interscience Publishers.

Boolootian, R.A., A.C. Giese, J.S. Tucker and A. Farmanfarmaian. 1959. A contribution to the biology of a deep sea echinoid *Allocentrotus fragilis* (Jackson). *Biological Bulletin* (Woods Hole, Massachusetts) 116: 362–72.

Boolootian, R.A., and D. Leighton. 1966. A key to the species of Ophiuroidea (brittle stars) of the Santa Monica Bay and adjacent areas. *Contributions in Science* (Natural History Museum of Los Angeles County) 93: 1–20.

Breen, P.A., W. Carolsfeld and K.L. Yamanaka. 1985. Social behavior of juvenile Red Sea Urchins, *Strongylocentrotus franciscanus*. *Journal of Experimental Marine Biology and Ecology* 92: 45–62.

Brehm, P., and J.G. Morin. 1977. Localization and characterization of luminescent cells in *Ophiopsila californicus* and *Amphipholis squamata* (Echinodermata: Ophiuroidea). *Biological Bulletin* (Woods Hole, Massachusetts) 152: 12–25.

Brusca, R.D., and G.J. Brusca. 1990. *Invertebrates*. Sunderland, Massachusetts: Sinauer Associates.

Byrne, M. 1994. Ophiuroidea. In *Microscopic Anatomy of Invertebrates*, edited by F.W. Harrison and F.S. Chia. Toronto: Wiley-Liss.

Byrne, M., and A.R. Fontaine. 1981. The feeding behaviour of *Florometra serratissima* (Echinodermata: Crinoidea). *Canadian Journal of Zoology* 59: 11–18.

– – –. 1983. Morphology and function of the tube feet of *Florometra serratissima* (Echinodermata: Crinoidea). *Zoomorphology* (Berlin) 102: 175–87.

Cavey, M.J., and K. Markel. 1994. Echinoidea. In *Microscopic Anatomy of Invertebrates*, edited by F.W. Harrison and F.S. Chia. Toronto: Wiley-Liss.

Chia, F.S. 1969a. Histology of the pedicellariae of the sand dollar, *Dendraster excentricus* (Echinodermata). *Journal of Zoology* 157: 503–07.

– – –. 1969b. Some observations on the locomotion and feeding of the sand dollar, *Dendraster excentricus* (Eschscholtz). *Journal of Experimental Marine Biology and Ecology* 3: 162–70.

Clark, A.H. 1907. Descriptions of new species of recent unstalked crinoids from the North Pacific Ocean. *Proceedings of the United States National Museum* 33: 69–84.

– – –. 1915. A monograph of the existing crinoids: Vol. 1, The Comatulids, Part 1. *United States National Museum Bulletin* 82: 1–795.

– – –. 1921. A monograph of the existing crinoids: Vol. 1, The Comatulids, Part 2. *United States National Museum Bulletin* 82: 1–795.

– – –. 1931. A monograph of the existing crinoids. Vol. 1, The Comatulids, Part 3. *United States National Museum Bulletin* 82: 1–816.

– – –. 1937. Crinoids of the Okhotsk and Japan Seas. *Exptl. Mers. U.S.S.R.* 23: 217–30.

– – –. 1941. A monograph of the existing crinoids. *Bulletin of the United States National Museum* 82: 1–603.

– – –. 1947. A monograph of the existing crinoids. *Bulletin of the United States National Museum* 82: 1–473.

– – –. 1950. A monograph of the existing crinoids. *Bulletin of the United States National Museum* 82: 1–383.

Clark, A.H., and A.M. Clark. 1967. A monograph of the existing crinoids. *Bulletin of the United States National Museum* 82: 1–860.

Clark, A.M. 1962. *Starfishes and Their Relations*. London: B.M.(N.H.).

– – –. 1970. Notes on the family Amphiuridae. *Bulletin of the British Museum of Natural History (Zoology)* 19: 1–81.

– – –. 1977. *Starfishes and Related Echinoderms*. Hong Kong: T.F.H. Publications (British Museum: Natural History).

Clark, H.L. 1911. North Pacific ophiurans in the collection of the United States National Museum. *Bulletin of the United States National Museum* 75: 1–302.

– – –. 1913. Article 8: Echinoderms from Lower California, with descriptions of new species. *Bulletin of the American Museum of Natural History* 32: 185–235.

— — —. 1940. Eastern Pacific expedition of the New York Zoological Society. 21. Notes on echinoderms from the west coast of Central America. *Zoologica* (New York) 25: 331–52.

— — —. 1948. A report on the Echini of the warmer eastern Pacific, based on the collections of the *Velero III*. *Allan Hancock Pacific Expeditions* 8: 1–352.

D'yakonov, A.M. 1949. Identification of the echinoderms of the Far Eastern seas. *Bulletin of the Pacific Scientific Research Institute of Marine Fisheries and Oceanography* 30: 1–130.

— — —. 1952. Echinoderms from abyssal depths in the waters around Kamtchatka. *Explorations of the Far East Seas of USSR* 3: 116–30.

— — —. 1958. Echinodermata, excluding Holothuroidea, collected by the Kurile-Sakhalin expedition in 1947–1949. *Explorations of the Far East Seas of USSR* 5: 271–357.

— — —. 1966. Class Ophiuroidea. In *Atlas of the Invertebrates of the Far Eastern Seas of the USSR*, edited by E.N. Pavlovskii. Jerusalem: Israel Program for Scientific Translations.

— — —. 1967. *Ophiuroids of the USSR Seas*. Jerusalem: Israel Program for Scientific Translations.

— — —. 1969. *Fauna of Russia and Adjacent Countries – Echinoidea*. Jerusalem: Israel Program for Scientific Translations.

de Ridder, C., and J.M. Lawrence. 1982. Food and feeding mechanisms: Echinoidea. In *Echinoderm Nutrition*, edited by M. Jangoux and J.M. Lawrence. Rotterdam: A.A. Balkema.

Deheyn, D., J. Mallefet and M. Jangoux. 1997. Intraspecific variations of bioluminescence in a polychromatic populations of *Amphipholis squamata* (Echinodermata: Ophiuroidea). *Journal of the Marine Biological Association of the United Kingdom* 77: 1213–22.

Durham, J.W. 1955. Classification of Clypeasteroid Echinoids. *University of California Publications in Geological Sciences* 31: 73–198.

Durham, J.W., H.B. Fell, A.G. Fischer, P.M. Kier, R.V. Melville, D.L. Pawson and C.D. Wagner. 1966. Echinoids. In *Treatise on Invertebrate Paleontology*, edited by R.C. Moore. Lawrence, Kansas: Geological Society of America and University of Kansas Press.

Durham, J.W., C.D. Wagner and D.P. Abbott. 1980. Echinoidea: The Sea Urchins. In *Intertidal Invertebrates of California*, edited by R.H. Morris, D.P. Abbott and E.C. Haderlie. Stanford, California: Stanford University Press.

Ebert, T.A. 1998. An analysis of the importance of Allee effects in management of the Red Sea Urchin *Strongylocentrotus franciscanus*. In *Echinoderms: San Francisco – Proceedings of the Ninth International Echinoderm Conference, San Francisco, California 5-9 August 1996*, edited by R. Mooi and M. Telford. Rotterdam: A.A. Balkema.

Ellers, O. 1985. Oral surface podial feeding in the sand dollar *Echinarachnius parma* (Lamarck). In *Echinodermata: Proceedings of the Fifth International Echinoderm Conference, Galway 24-29 September 1984*, edited by B.F. Keegan and B.D.S. O'Connor. Rotterdam: A.A. Balkema.

Ellers, O., and M. Telford. 1984. Collection of food by oral surface podia in the sand dollar, *Echinarachnius parma* (Lamarck). *Biological Bulletin* (Woods Hole, Massachusetts) 166: 574–82.

Emson, R.H., and J. Foote. 1979. Environmental tolerances and other adaptive features of two intertidal rock pool echinoderms. In *Proceedings of the European Colloquium on Echinoderms*, edited by M. Jangoux. Rotterdam: A.A. Balkema.

Emson, R.H., and I.C. Wilkie. 1982. The arm-coiling response of *Amphipholis squamata* (Delle Chiaje). In *Echinoderms: Proceedings of the International Conference, Tampa Bay*, edited by J.M. Lawrence. Rotterdam: A.A. Balkema.

Estes, J.A., D.O. Duggins and G.B. Rathbun. 1989. The ecology of extinctions in kelp forest communities. *Conservation Biology* 3: 252–64.

Falk-Petersen, I.-B. 1983. Light and electron microscopical studies of the embryonic development of *Strongylocentrotus pallidus* (G.O. Sars). *Sarsia* 68: 9–20.

Fell, F.J. 1982. Echinodermata. In *Synopses and Classification of Living Organisms*, edited by S.P. Parker. Toronto: McGraw-Hill.

Fell, H.B. 1960. *Synoptic Keys to the Genera of Ophiuroidea*. Zoological Publication 26. Wellington, Australia: Victoria University.

Fontaine, A.R., and P. Lambert. 1976. The fine structure of the sperm of a holothurian and an ophiuroid. *Journal of Morphology* 148: 209–25.

Fujita, T., and S. Ohta. 1988. Photographic observations of the lifestyle of a deep-sea ophiuroid *Asteronyx loveni* (Echinodermata). *Deep-Sea Research, Part A, Oceanographic Research Papers* 35: 2029–43.

— — —. 1989. Spacial structure within a dense bed of the brittle star *Ophiura sarsi* (Ophiuroidea: Echinodermata) in the bathyal zone off Otsuchi, northeastern Japan. *Journal of the Oceanographic Society of Japan* 45: 289–300.

Ghiold, J. 1982. The feeding and burrowing mechanism in the sand dollar *Echinarachnius parma*. *Bulletin Mt Desert Island Biological Laboratory* 21: 21.

Giese, A.C. 1961. Further studies on *Allocentrotus fragilis*, a deep sea echinoid. *Biological Bulletin* (Woods Hole, Massachusetts) 121: 141–50.

Giese, A.C., J.S. Pearse and V.B. Pearse. 1991. *Reproduction of Marine Invertebrates: Echinoderms and Lophophorates*. Pacific Grove, California: Boxwood Press.

Gilkinson, K.D., J.-M. Gagnon and D.C. Schneider. 1988. The sea urchin *Strongylocentrotus pallidus* (G.O. Sars) on the Grand Bank of Newfoundland. In *Echinoderm Biology*, edited by R.D. Burke, P.V. Mladenov, P. Lambert and R.L. Parsley. Rotterdam: A.A. Balkema.

Grainger, E.H. 1955. Echinoderms of Ungava Bay, Hudson Strait, Frobisher

Bay and Cumberland Sound. *Journal of the Fisheries Research Board of Canada* 12: 899–916.

Harbo, R. 1999. *Whelks to Whales: Coastal Marine Life of the Pacific Northwest.* Madeira Park, BC: Harbour Publishing.

Harold, A.S., and M. Telford. 1982. Substrate preference and distribution of the northern sand dollar, *Echinarachnius parma*. In *Echinoderms: Proceedings of the International Conference, Tampa Bay, 14-17 September 1981,* edited by J.M. Lawrence. Rotterdam: Balkema.

Hart, M.W. 1996. Evolutionary loss of larval feeding: development, form and function in a facultatively feeding larva, *Brisaster latifrons. Evolution* 50: 174–87.

Heinzeller, T., and U. Welsch. 1994. Crinoidea. In *Microscopic Anatomy of Invertebrates,* edited by F.W. Harrison and F.S. Chia. Toronto: Wiley-Liss.

Hendler, G. 1975. Adaptational significance of the patterns of ophiuroid development. *American Zoologist* 15: 691–715.

———. 1996a. Class Crinoidea. In *Taxonomic Atlas of the Benthic Fauna of the Santa Maria Basin and the Western Santa Barbara Channel,* edited by J.A. Blake, P.H. Scott and A. Lissner. Santa Barbara, California: Santa Barbara Museum of Natural History.

———. 1996b. Class Ophiuroidea. In *Taxonomic Atlas of the Benthic Fauna of the Santa Maria Basin and the Western Santa Barbara Channel,* edited by J.A. Blake, P.H. Scott and A. Lissner. Santa Barbara, California: Santa Barbara Museum of Natural History.

———. 2007. Ophiuroidea. In *The Light & Smith Manual: Intertidal Invertebrates from Central California to Oregon,* edited by J.T. Carlton. Berkeley and Los Angeles: University of California Press.

Hood, S., and R. Mooi. 1998. Taxonomy and phylogenetics of extant *Brisaster* (Echinoidea: Spatangoida). In *Echinoderms: San Francisco. Proceedings of the Ninth International Echinoderm Conference, San Francisco, California, USA, 5-9 August 1996,* edited by R. Mooi and M. Telford. Rotterdam: A.A. Balkema.

Hyman, L.H. 1955. *The Invertebrates: Echinodermata – The Coelomate Bilateria,* vol. 4. Toronto, Ont.: McGraw-Hill Book Co.

Imaoka, T., S. Irimura, T. Okutani, C. Oguro, T. Oji, M. Shigei and H. Horikawa. 1990. *Echinoderms from Continental Shelf and Slope Around Japan.* Tokyo: Japan Fisheries Resource Conservation Association.

Ivanov, B.G. 1964. Quantitative distribution of the echinoderms on the shelf of the East Bering Sea. *Trans. Inst. Mar. Fish. U.S.S.R.* 49: 123–40.

Jackson, R.T. 1912. Phylogeny of the Echini, with a revision of Palaeozoic species. *Memoirs of the Boston Society of Natural History* 7: 199–235.

Jangoux, M. 1982. Digestive systems: Ophiuroidea. In *Echinoderm nutrition,* edited by M. Jangoux and J.M. Lawrence. Rotterdam: A.A. Balkema.

Jensen, M. 1974. The Strongylocentrotidae (Echinoidea), a morphologic and systematic study. *Sarsia* 57: 113–48.

Kier, P.M. 1987. Class Echinoidea. In *Fossil Invertebrates*, edited by R.S. Boardman, A.H. Cheetham and A.J. Rowell. Palo Alto, California: Blackwell Scientific Publications.

Kozloff, E.N. 1983. *Seashore Life of the Northern Pacific Coast*. Seattle: University of Washington Press.

———. 1987. *Marine Invertebrates of the Pacific Northwest*. Seattle: University of Washington Press.

Kyte, M.A. 1969. A synopsis and key to the recent Ophiuroidea of Washington State and southern British Columbia. *Journal of the Fisheries Research Board of Canada* 26: 1727–41.

———. 1987. *Stegophiura ponderosa* (Lyman), new combination, and *Amphiophiura vemae* and *Homophiura nexila*, new species (Echinodermata: Ophiuroidea) from the R/V *Vema* collections. *Proceedings of the Biological Society of Washington* 100: 249–56.

Labarbera, M. 1978. Particle capture by a Pacific brittle star: experimental test of the aerosol suspension feeding model. *Science* 201: 1147–49.

Lamb, A., and B.P. Hanby. 2005. *Marine Life of the Pacific Northwest: A Photographic Encyclopedia of Invertebrates, Seaweeds and Selected Fishes*. Madeira Park, BC: Harbour Publishing.

Laur, D.R., A.W. Ebeling and D.C. Reed. 1986. Experimental evaluations of substrate types and barriers to sea urchin (*Strongylocentrotus* spp.) movement. *Marine Biology* (Berlin) 93: 209–16.

Lawrence, J. 1987. *A Functional Biology of Echinoderms*. London and Sydney: Croom Helm.

Lawrence, J.M. 1975. On the relationships between marine plants and sea urchins. *Oceanography and Marine Biology, Annual Review* 13: 213–86.

Litvinova, N.M. 1981. Behaviour of *Ophiopholis aculeata* (Ophiuroidea) in the time of reproduction. *Zoologicheskii Zhurnal* 60: 942–45.

Litvinova, N.M., and M.N. Sokolova. 1971. Feeding of deep-sea ophiuroids of the genus *Amphiophiura*. *Oceanology* 11: 240–47.

Lyman, T. 1860. Descriptions of new Ophiuridae, belonging to the Smithsonian Institution and to the Museum of Comparative Zoology at Cambridge. *Proceedings of the Boston Society of Natural History* 7: 193–205.

———. 1865. *Illustrated Catalogue of the Museum of Comparative Zoology at Harvard College*. No. 1: *Ophiuridae and Astrophytidae*. Cambridge: Legislature of Massachusetts.

———. 1878. Ophiuridae and Astrophytidae of the voyage of HMS *Challenger* under Prof. Sir Wyville Thomson, F.R.S. Part 1, no. 7. *Bulletin of the Museum of Comparative Zoology* (Harvard) 5: 65–168.

———. 1879. Ophiuridae and Astrophytidae of the exploring voyage of HMS *Challenger* under Prof. Sir Wyville Thomson, F.R.S. Part 2, no. 2. *Bulletin of the Museum of Comparative Zoology* (Harvard) 6: 17-83.

MacGinitie, G.E., and N. MacGinitie. 1968. *Natural History of Marine Animals*. Toronto: McGraw-Hill.

Maier, D., and P. Roe. 1983. Preliminary investigations of burrow defense and intraspecific aggression in the sea urchin, *Strongylocentrotus purpuratus*. *Pacific Science* 37: 145–49.

Matsumoto, H. 1917. A monograph of Japanese Ophiuroidea, arranged according to a new classification. *Journal of the College of Science, Imperial University, Tokyo* 38: 1–408.

McCauley, J.E. 1967. Status of the heart urchin, *Brisaster latifrons*. *Journal of the Fisheries Research Board of Canada* 24: 1377–84.

McCauley, J.E., and A.G. Carey, Jr. 1967. Echinoidea of Oregon. *Journal of the Fisheries Research Board of Canada* 24: 1385–1401.

McClendon, J.F. 1909. The ophiurans of the San Diego region. *University of California Publications in Zoology* 6: 33–64.

McCloskey, L.R. 1970. A new species of *Dulichia* (Amphiphoda, Podoceridae) commensal with a sea urchin. *Pacific Science* 24: 90–98.

McDaniel, N. 1976. The feather star. *Pacific Diver* 2: 24–27.

McEdward, L.R., S.F. Carson and F.S. Chia. 1988. Energetic content of eggs, larvae, and juveniles of *Florometra serratissima* and the implications for the evolution of crinoid life histories. *Invertebrate Reproduction and Development* 13: 9–22.

Messing, C.G., and C.M. White. 2001. A revision of the Zenometridae (new rank) (Echinodermata, Crinoidea, Comatulidina). *Zoologica Scripta* 30: 159–80.

Mironov, A.N. 1976. Deep-sea urchins of the northern Pacific. *Trudy Instituta Okeanologii, Akademiya Nauk SSSR* 99: 140–64.

Mladenov, P.V. 1981. *Development and reproductive biology of the feather star Florometra serratissima (Echinodermata: Crinoidea)*. Ph.D. Thesis, University of Alberta, Edmonton.

— — —. 1982. Development and larval behaviour in the feather star *Florometra serratissima* (A.H. Clarke). In *Echinoderms: Proceedings of the International Conference, Tampa Bay, 14-17 September 1981*, edited by J.M. Lawrence. Rotterdam: Balkema.

— — —. 1983. Rate of arm regeneration and potential causes of arm loss in the feather star *Florometra serratissima* (Echinodermata: Crinoidea). *Canadian Journal of Zoology* 61: 2873–79.

— — —. 1986. Reproductive biology of the feather star *Florometra serratissima*: gonadal structure, breeding pattern, and periodicity of ovulation. *Canadian Journal of Zoology* 64: 1642–51.

Mladenov, P.V., and F.S. Chia. 1983. Development, settling behaviour, metamorphosis and pentacrinoid feeding and growth of the feather star *Florometra serratissima*. *Marine Biology* (Berlin) 73: 309–23.

Moitoza, D.J., and D.W. Phillips. 1979. Prey defense, predator preference and non-random diet: the interactions between *Pycnopodia helianthoides* and two species of sea urchins. *Marine Biology* (Berlin) 53: 299–304.

Mooi, R. 1986a. Non-respiratory podia of clypeasteroids (Echinodermata, Echinoides). 1: Functional anatomy. *Zoomorphology* 106: 21–30.

— — —. 1986b. Non-respiratory podia of clypeasteroids (Echinodermata, Echinoides). 2: Diversity. *Zoomorphology* 106: 75–90.

— — —. 1986c. Structure and function of clypeasteroid miliary spines (Echinodermata, Echinoides). *Zoomorphology* 106: 212–23.

— — —. 1989. Living and fossil genera of the Clypeasteroida (Echinoidea: Echinodermata): an illustrated key and annotated checklist. *Smithsonian Contributions to Zoology* 488: 1–51.

— — —. 1997. Sand dollars of the genus *Dendraster* (Echinoidea: Clypeastroida): phylogenetic systematics, heterochrony, and distribution of extant species. *Bulletin of Marine Science* 61: 343–75.

Mooi, R., and B. David. 1996. Phylogenetic analysis of extreme morphologies: deep-sea holasteroid echinoids. *Journal of Natural History* 30: 913–53.

Mooi, R., and M. Telford. 1982. The feeding mechanism of the sand dollar *Echinarachnius parma* (Lamarck). In *Echinoderms: Proceedings of the International Conference, Tampa Bay, 14-17 September 1981*, edited by J.M. Lawrence. Rotterdam: A.A. Balkema.

Moore, A.R. 1959a. On the embryonic development of the sea urchin *Allocentrotus fragilis*. *Biological Bulletin* (Woods Hole, Massachusetts) 117: 492–96.

— — —. 1959b. Some effects of temperature on the embryonic development in the sea urchin *Allocentrotus fragilis*. *Biological Bulletin* (Woods Hole, Massachusetts) 117: 150–53.

Moore, R.C., ed. 1966. *Treatise on Invertebrate Paleontology*. 24 volumes. New York: The Geological Society of America; Lawrence: The University of Kansas Press.

Morris, R.H., D.P. Abbott and E.C. Haderlie. 1980. *Intertidal Invertebrates of California*. Stanford, California: Stanford University Press.

Mortensen, T. 1910. The echinoidea of the Swedish south-polar expedition. *Wissenschaftliche Ergebnisse der Schwedischen Südpolar-Expedition, 1901–1903* 6: 114.

— — —. 1921. *Studies of the Development and Larval Forms of Echinoderms*. Copenhagen: Carlsberg Fund, G.E.C. Gad.

— — —. 1928. *Cidaroida*. Vol. 1 of *A Monograph of the Echinoidea*. Copenhagen: C.A. Reitzel.

— — —. 1942. New Echinoidea (Camarodonta). *Videnskabelige Meddelelser fra Dansk Naturhistorisk Forening i Khobenhavn* 106: 225–32.

— — —. 1948. Clypeastroida: Clypeastridae, Arachnoididae, Fibulariidae, Laganidae and Scutellidae. Vol. 4, part 2 of *A Monograph of the Echinoidea*. Copenhagen: C.A.Reitzel.

— — —. 1951. Spatangoida II: Amphisternata. II. Spatangidae, Loveniidae, Pericosmidae, Scizasteridae, Brissidae. Vol. 5, part 2 of *A Monograph of the Echinoidea*. Copenhagen: C.A. Reitzel.

— — —. 1977. *Handbook of the Echinoderms of the British Isles*. Reprint of 1927 edition. Rotterdam: Dr W. Backhuys, Uitgever.

Nakano, H., T. Hibino, T. Oji, Y. Hara and S. Amemiya. 2003. Larval stages of a living sea lily (stalked crinoid echinoderm). *Nature* (London) 421: 158–60.

Neve, R.A., and G.A. Howard. 1970. Carotenoids of the crinoid *Florometra serratissima. Occ. Publs. Inst. Mar. Sci. Univ. Alaska* 1: 549–62.

Nichols, D. 1969. *Echinoderms.* London: Hutchinson University Library.

Nichols, F.H. 1975. Dynamics and energetics of three deposit-feeding benthic invertebrate populations in Puget Sound, Washington. *Ecological Monographs* 45: 57–82.

Nielsen, E. 1932. Ophiurans from the Gulf of Panama, California and Strait of Georgia. *Videnskabelige Meddelelser fra Dansk Naturhistorisk Forening i Khobenhavn* 91: 241–336.

Palumbi, S.R., and A.C. Wilson. 1990. Mitochondrial DNA diversity in the sea urchins *Strongylocentrotus purpuratus* and *S. droebachiensis. Evolution* 44: 403–15.

Patent, D.H. 1969. The reproductive cycle of *Gorgonocephalus caryi* (Echinodermata, Ophiuroidea). *Biological Bulletin* (Woods Hole, Massachusetts) 136: 241–52.

— — —. 1970. Life history of the basket star, *Gorgonocephalus eucnemis* (Muller and Troschel) (Echinodermata, Ophiuroidea). *Ophelia* 8: 145–60.

Paterson, G.L.J. 1985. The deep-sea Ophiuroidea of the North Atlantic Ocean. *Bulletin of the British Museum of Natural History (Zoology)* 49: 1–162.

Rader, D.N. 1982. Orthonectid parasitism: effects on the ophiuroid, *Amphipholis squamata.* In *Echinoderms: Proceedings of the International Conference, Tampa Bay, 14-17 September 1981,* edited by J.M. Lawrence. Rotterdam: A.A. Balkema.

Ricketts, E.F., and J. Calvin. 1964. *Between Pacific Tides.* Stanford, California: Stanford University Press.

Ricketts, E.F., J. Calvin and J.W. Hedgpeth. 1985. *Between Pacific Tides.* Stanford, California: Stanford University Press.

Roller, R.A., and W.B. Stickle. 1985. Effects of salinity on larval tolerance and early developmental rates of four species of echinoderms. *Canadian Journal of Zoology* 63: 1531–38.

Rumrill, S.S. 1982. Contrasting reproductive patterns among brittle stars from Monterey Bay, California. In *Echinoderms: Proceedings of the International Conference, Tampa Bay, 14-17 September 1981,* edited by J.M. Lawrence. Rotterdam: A.A. Balkema.

Rumrill, S.S., and J.S. Pearse. 1985. Contrasting reproductive periodicities among north eastern Pacific ophiuroids. In *Echinodermata: Proceedings of the Fifth International Echinoderm Conference, Galway, 24-29 September 1984,* edited by B.F. Keegan and B.D.S. O'Connor. Rotterdam: A.A. Balkema.

Salazar, M.H. 1970. Phototaxis in the deep-sea urchin *Allocentrotus fragilis* (Jackson). *Journal of Experimental Marine Biology and Ecology* 5: 254–64.

Schiff, K.C., and M. Bergen. 1995. Impact of Wastewater on Reproduction of *Amphiodia urtica*. *Southern California Coastal Water Research Project 1994-95 Annual Report*. Los Angeles: Southern California Coastal Water Research Project.

Scouras, A., K. Beckenbach, A. Arndt and M.J. Smith. 2004. Complete mitochondrial genome DNA sequence for two ophiuroids and a holothuroid: the utility of protein gene sequence and gene maps in the analyses of deep deuterostome phylogeny. *Molecular Phylogenetics and Evolution* 31: 50–65.

Scouras, A., and M.J. Smith. 2001. A novel mitochondrial gene order in the crinoid echinoderm *Florometra serratissima*. *Molecular Biology and Evolution* 18: 61–73.

Sept, J.D. 1999. *The Beachcomber's Guide to Seashore Life in the Pacific Northwest*. Madeira Park, BC: Harbour Publishing.

Shaw, G.D., and A.R. Fontaine. 1990. The locomotion of the comatulid *Florometra serratissima* (Echinodermata: crinoidea) and its adaptive significance. *Canadian Journal of Zoology* 68: 942–50.

Shinn, G.L. 1986. Spontaneous hatching of *Fallocohospes inchoatus*, an umagillid flatworm from the northeastern Pacific crinoid *Florometra serratissima*. *Canadian Journal of Zoology* 64: 2068–71.

Smith, A.B., G.L.J. Paterson and B. Lafay. 1995. Ophiuroid phylogeny and higher taxonomy: morphological, molecular and palaeontological perspectives. *Zoological Journal of the Linnean Society* 114: 213–43.

Smith, C.R., and S.C. Hamilton. 1983. Epibenthic megafauna of a bathyal basin off southern California: patterns of abundance, biomass and dispersion. *Deep-Sea Research* 30: 907–28.

Spencer, W.K., and C.W. Wright. 1966. Asterozoans. In *Treatise on Invertebrate Paleontology*, Part U: *Echinodermata 3*, edited by R.C. Moore. Lawrence: University Press of Kansas.

Stancyk, S.E., C. Muir and T. Fujita. 1998. Predation behavior on swimming organisms by *Ophiura sarsii*. In *Echinoderms: San Francisco – Proceedings of the Ninth International Echinoderm Conference, San Francisco, California, 5-9 August 1996*, edited by R. Mooi and M. Telford. Rotterdam: A.A. Balkema.

Steimle, F.W. 1990. Population dynamics, growth, and production estimates of the sand dollar *Echinarachnius parma*. *Fishery Bulletin of the Fish and Wildlife Service* 88: 179–89.

Stien, A. 1999. Effects of the parasitic nematode *Echinomermella matsi* on growth and survival of its host, the sea urchin *Strongylocentrotus droebachiensis*. *Canadian Journal of Zoology* 77: 139–47.

Strathmann, M.F. 1987. *Reproduction and Development of Marine Invertebrates of the Northern Pacific Coast: Data and Methods for the Study of Eggs, Embryos and Larvae*. Seattle: University of Washington Press.

Strathmann, M.F., and S.S. Rumrill. 1987. Phylum Echinodermata, Class Ophiuroidea. In *Reproduction and Development of Marine Invertebrates of* the *Northern Pacific Coast*, edited by M.F. Strathmann. Seattle: University of Washington Press.

Strathmann, R.R. 1971. The feeding behavior of planktotrophic echinoderm larvae: mechanisms, regulation, and rates of suspension feeding. *Journal of Experimental Marine Biology and Ecology* 6: 109–60.

———. 1975. Larval feeding in echinoderms. *American Zoologist* 15: 717–30.

———. 1978. Length of pelagic period in echinoderms with feeding larvae from the northeast Pacific. *Journal of Experimental Marine Biology and Ecology* 34: 23–27.

———. 1979. Echinoid larvae from the northeast Pacific (with a key and comment on an unusual type of planktotrophic development). *Canadian Journal of Zoology* 57: 610–16.

Sumich, J.L. 1973. Growth of a sea urchin, *Allocentrotus fragilis*, at different depths off the Oregon coast. *Bulletin of the Ecological Society of America* 52: 20.

Sumich, J.L., and J.E. McCauley. 1973. Growth of a sea urchin, *Allocentrotus fragilis*, off the Oregon Coast. *Pacific Science* 27: 156–67.

Summers, A.C., and J. Nybakken. 2000. Brittle star distribution patterns and population densities on the continental slope off central California (Echinodermata: Ophiuroidea). *Deep Sea Research* Part II: *Tropical Studies in Oceanography* 47: 1107–37.

Summers, R.G., B.L. Hylander, L.H. Colwin and A.L. Colwin. 1975. The functional anatomy of the echinoderm spermatozoon and its interaction with the egg at fertilization. *American Zoologist* 15: 523–51.

Swan, E.F. 1962. Evidence suggesting the existence of two species of *Strongylocentrotus* (Echinoidea) in the northwest Atlantic. *Canadian Journal of Zoology* 40: 1211–22.

Talmadge, R.R. 1976. A double echinoderm. *Of Sea and Shore* 7: 171.

Thompson, B., G. Jones, J. Laughlin and D. Tsukada. 1987. Distribution, abundance, and size-composition of echinoids from basin slopes off southern California. *Bulletin of the Southern California Academy of Sciences* 86: 113–25.

Timko, P.L. 1976. Sand dollars as suspension feeders: a new description of feeding in *Dendraster excentricus*. *Biological Bulletin* (Woods Hole, Massachusetts) 151: 247–59.

Vadas, R.L., R.W. Elner, P.E. Garwood and I.G. Babb. 1986. Experimental evaluation of aggregation behaviour in the sea urchin *Strongylocentrotus droebachiensis*: a reinterpretation. *Marine Biology* (Berlin) 90: 433–48.

Warner, G. 1982. Food and feeding mechanisms: Ophiuroidea. In *Echinoderm Nutrition*, edited by M. Jangoux and J.M. Lawrence. Rotterdam: A.A. Balkema.

Webster, S.K. 1975. Oxygen consumption in echinoderms from several geographical locations, with particular reference to the Echinoidea. *Biological Bulletin* (Woods Hole, Massachusetts) 148: 157–64.

Westervelt, C.A.J., and E.N. Kozloff. 1992. Two new species of *Syndesmis* (Turbellaria: Neorhabdocoela: Umagillidae) from the sea urchins *Strongylocentrotus droebachiensis* and *Allocentrotus fragilis*. *Cahiers de Biologie Marine* 33: 115–24.

Wilkie, I.C. 1988. Design for disaster: The ophiuroid intervertebral ligament as a typical mutable collagenous structure. In *Echinoderm Biology*, edited by R.D. Burke, P.V. Mladenov, P. Lambert and R.L. Parsley. Rotterdam: A.A. Balkema.

ACKNOWLEDGEMENTS

The authors thank the Royal British Columbia Museum for supporting Philip Lambert during the compilation and writing of this book. Thanks also to Gerry Truscott for editing and shepherding this volume through the publishing process. We thank the following people and organizations for making specimens available: D. Pawson and Cynthia Ahearn (National Museum of Natural History), R. Woolacott (Museum of Comparative Zoology), D. Abbott (Stanford University), F. Ziesenhenne (Allan Hancock Foundation), curators at the Zoological Museum in Copenhagen, Bob van Syoc, the California Academy of Sciences, Bruce Wing, the Auke Bay Marine Lab and the National Museum of Canada. Many thanks also to: Dr Michael Kyte for assistance with identifications; Dr Jim Boutillier, Department of Fisheries and Oceans, for inviting us on several deep-sea collecting trips and providing the opportunity to obtain images and specimens; and Myriam Preker (Simon Fraser University), who worked with William Austin on ophiuroid collections from the Pacific Biological Station, Bamfield Marine Station and Simon Fraser University. William Austin's passion in brittle star biology and ecology was fostered through the mentorship of Don Abbott at Hopkins Marine Station, Stanford University, and Gunner Thorson at the Marine Biological Laboratory, University of Copenhagen. We are especially grateful to Dr Rich Mooi and Dr Gordon Hendler for thoughtful and constructive criticisms of the manuscript.

Brittle Stars, Sea Urchins and Feather Stars
of British Columbia, Southeast Alaska and Puget Sound

Edited, designed and typeset by Gerry Truscott, RBCM.
Set in Palatino (body 10/12) and Optima (captions 9/11).
Cover design by Chris Tyrrell, RBCM.
Printed in Canada by Hignell Book Printing.

Drawings:
• Philip Lambert (RBCM ©): figures 2-24, 26–72.
• Dana Bean ©: figure 25.
• Gerald Luxton (RBCM ©): figure 1.
• Rick Pawlas (RBCM ©): map on page 2.

Photographs:
• William Austin ©: figure 50, C-16, C-17, C-20, C-21, C-23, C-24, C-26, C-27, C-28, C-30, C-32, C-33, C-34, C-37, C38.
• Philip Lambert (RBCM ©): back cover, figures 9 and 20, C-8, C-9, C-11, C-13, C-22, C-29, C-36, C-39, C-40;
• Philip Lambert ©: C-3, C-10.
• Brent Cooke (RBCM ©): front cover, C-6, C-7, C-12, C-14, C-15.
• Jim Boutillier ©: C-18, C-19, C-25, C-35.
• Jim Cosgrove ©: C-2, C-4, C-5.
• Olympic Coast National Marine Sanctuary ©: C-1.
• Keoki Stender ©: C-31.

INDEX

Bold indicates the description for the species, genus or family.
"C" designates a colour photograph.